Longwall Mining at Shallow Depth of Cover In India

Longwall Mining at Shallow Depth of Cover In India

Tapan Kumar Mozumdar;
Biswanath Pan

PARTRIDGE
A Penguin Random House Company

To order additional copies of this book, contact
Partridge India
000 800 10062 62
orders.india@partridgepublishing.com

www.partridgepublishing.com/india

Contents

PREFACE

The feasibility of extraction by Long wall method at shallow depth of cover was not considered earlier as the majority of the topmost seams in almost all the Coal Fields in India were developed by Board & Pillar method. Moreover, the earlier experience of extraction by Long wall method in two (2) to three (3) mines with H/T ratio less than ten (10) (where H is hardcover and T is extraction thickness) was not favourable as extraction could not be completed due to irregular caving even after induced blasting for surface was resorted to. It was therefore decided that Long wall panel should not be designed at shallow depth of cover with H/T ratio less than ten (10) and powered support capacity with effective mean load density of less than 80 T/sq.m.

The extraction by Long wall method with Chinese powered supports for use under geo-mining conditions existing in the mines under South Eastern Coalfields Ltd (SECL) was planned and 450 tonnes Chinese powered supports recommended by the Central Mining Research Institute (CMRI) was approved by the Directorate General of Mines Safety (DGMS). The rated load density of 450 Tonnes capacity support before cut was 80.3 T/m^2 against recommended 66 T/m^2 and after cut was 69.2 T/m^2 against recommended 57 T/m^2.

Extraction in one panel(P-1) in a mine under South Eastern Coal Field Ltd. was started on 11-5-98 and the first main fall occurred when the face advanced to 80m with an exposed area of about 12,000sq.m.and then 16 supports in the mid zone collared. After recovery of the supports, face operations continued. The second main fall and weighting occurred when the face advanced to 160m. with an area of exposure 24000 sq.m. A number of powered supports started bleeding and maximum cumulative convergence *of one* of the supports was found to be 630mm. The peak leg pressure in some of the supports located in the central zone showed above 40 MPa resulting in collaring of 13 supports. Crack was observed in the roof in between the face and tip of the front canopy. On scrutiny of the yield valves of the supports it was observed that a number of yield valves failed. Face had to be kept stopped for 46 days.

A detailed examination of individual supports revealed the failure of the following components.

(i)   13 supports collared & canopy got damaged
(ii) Bulging of sector rings of powered support leg.
(iii) Bending of inner tubes of legs
(iv) Leakage in non-returns valve.
(v)   Mal-functioning pressure gauges

Careful scrutiny of the failed components by a group of experts revealed some defects and these components were replaced. The management did not want to scrap the supports and purchase new supports with higher capacity. Therefore, the group of experts took the decisions to continue with the same chock shield supports up rate to 513 Tonnes by taking the following actions:

i)   By increasing the setting in bleed valves to increase the yield pressure from 35 MPa to 40MPa.
ii) To adjust the mechanical extension of the support to reduce overhang.
iii) To install positive set values in the leg circuits to ensure desired high setting load.
iv) To introduce induced caving by deep hole blasting from surface in order to reduce overhang to 30m and excessive loading on supports.

The face was again started after implementation of most of the above decisions. Deep Hole Blasting from surface was started at 178m travel to induce caving when the face was at 191 m travel and blasting interval was maintained at 15 m and later on increased to 30m and again reduced to 15m till the end up to 679 m travel.

The impact of induced caving on some important parameters (as listed below) was observed for successful operations and the results are presented in brief.

1) *Loading on supports* –increased loading on supports.
2) *Periodicity of weighting-* reduction in periodicity of weighting beyond 30m and increase of regularity in weighting between 10-20m.

3) *Convergence-* When the blasting interval was 15m, cumulative convergence was in the range of 25mm to 108mm but when the interval rose to 30m, cumulative convergence increased to the range of 40mm to 410mm.

4) *Subsidence* – the magnitude of negative angle of draw was reduced thereby indicating reduced overhang on face.

Presence of hard and massive roof in the caving zone for seams to be extracted by underground method was observed in the mines under South Eastern Coalfields Ltd (SECL), Singareni Coalfields Ltd. (SECL) and some other Coalfields. In the present study it was established that induced caving by blasting from surface could solve some of the problems at shallow depth of cover. But extraction of the seams occurring at greater depth with difficult to cave in roof under normal circumstances has to be planned by induced caving from underground or hydro fracturing or by combination of both. Attempt has already been started for induced caving from underground but the technology has to be better designed and improved for successful operation. The experience of the study would be of immense help in planning for extraction of coal seams in India at shallow depth of hard cover.

1.0 ABSTRACT

Long wall projects for Balrampur, New Kumda *&* Rajendra underground (U/G) mine with powered supports manufactured by CME I&E, CHINA were approved in DEC 94. The Central Mining Research Institute(CMRI) assessed the support capacity for these projects based on experiences of shallow cover long wall operations made elsewhere in India, which was also approved by the Directorate General of Mines Safety (DGMS) for implementation. First L/W panel was started in Balrampur followed by in Rajendra U/G mine.

This paper deals with the case study of two completed long wall panels of Balrampur & one of Rajendra U/G mine working at shallow depth of cover.

The strata control observations during operation of long wall panels had established that the support capacity was marginally adequate under normal circumstances but inadequate to provide the desired support resistance during main (major) weighting and also during periodic weightings. In absence of any choice for use of higher capacity support, the supports manufactured by CME were up rated as far as possible under the design parameters and the same were used in the first panel and the other panels. Along with marginally up rating of supports, induced caving by blasting from surface was resorted to experimentally.

The observations made after introduction of induced caving had established that loading on supports, due to development of discontinuous subsidence trough associated with stepped subsidence had increased. These had resulted in more regular occurrence of weightings and reduction of convergence due to more filling up of the voids/goaves. The extent of caving zone initially influenced by the physico-mechanical properties of overlying rock was not always immediately activated but the induced caving had reduced the overhang which reflected in reduced convergence. The contribution of induced caving ultimately proved to be one of the important operations for success of long wall mining under the existing conditions and the practice had, therefore;

been made compulsory for all the long wall operations where the same capacity supports would be used under the same geo-mining conditions. However, the failure in achieving the desired results in strata management by induced caving in all the panels was due to absence of optimization in use of explosives and good blast design. In the coming days, the blast design needed optimization to achieve full benefits from such operations. The strata control observations had also established that loading on supports was comparatively more when the Hard cover to height of extraction ratio (H/t ratio) approached towards 10 and the support loading in this locale was much higher than that at the mine under Raniganj Coalfields worked at lesser H/t ratio (i.e. less than 10). As a result, the support capacity assessment based on the experience of the mine under Raniganj Coalfields was found to be inadequate.

The management of men and machine established that rate of face advance had direct relation with the loads & convergence on the support as established universally but on some occasions when the dynamic loading was more, collaring and damage of supports were observed. Good management of men and machines in all the panels reflected in daily production rate which was much higher than the planned production. However, higher rate of face advance (9 to 10 m) per day developed problems in control of roof due to higher loading and therefore advance of face had to be restricted to an average 6/7 m per day. Higher production rate improved Output per Man Shift (OMS) & profitability. The experiences in the operations of number of panels also highlighted the importance of detailed geo- technical studies of overlying rock at the planning stage and choice of appropriate capacity of support for success of long wall operations even at a shallow depth of cover. The subsidence as expected was more at lower H/t ratio as the overlying strata did not behave as continuous trough but as discontinuous trough associated with stepped subsidence. The travelling angle of draw also gave sufficient indication of hanging strata and provided guidance for immediate induced caving for reduction of overhang in underground. Review of the strata control management of the long wall operations at Jhanjra Project in the Raniganj coal fields under Eastern Coalfield Ltd.(ECL) was also made.

Keywords: Induced Caving, Discontinuous Subsidence, Stepped Subsidence, Optimization, H/T ratio

2.0 INTRODUCTION

Coal reserves up to 100 m depth of cover were so far been mined mainly by underground board & pillar or by opencast method. During the last two decades the mechanized opencast mining took a lead by contributing to about 75% of total coal production both from virgin and developed seams. In case of thick seam, use of blasting gallery method for development on Board and Pillar was, however, tried in limited areas but for seams highly susceptible to spontaneous heating, the method proved to be unsafe due to possibility of outbreak of fire. The long wall at shallow depth of cover was not thought of earlier and was not tried at properties lying at a depth of cover less than 60 m.

In India, coal production by long wall was about 2.63 percent of underground production in the year 1996-97. The scope of long wall at shallow depth of cover was not considered in particular as the top most seams in almost all coalfields of India were already developed by Board &Pillar. Long wall at shallow depth of cover was tried at Bijuri u/g mine of SECL. The Longwall face could not be completed due to irregular caving even after induced blasting from surface. In recent past Jhanjra mine of Raniganj Coalfield Ltd.(RCF) was worked by longwall method at a depth of cover around 50m. Presently 3 mines of SECL namely Balrampur, New Kumda, Rajendra were being worked by longwall method at a shallow depth of cover around 50m.

3.0 REVIEW OF PAST EXPERIENCE

A number of long wall panels at Jhanjra Project with a depth of cover varying from 40 m to 60 m (hard cover from 16 m to 40 m) and thickness of seam varying from 3 m to 4 m were worked successfully (Ref Table 1,2,3). It was experienced that whenever H/t ratio (where H is the hard cover and t-thickness of extraction) was 10 or more there was no significant strata control problems. While working a panel with H/t ratio about 8, a lot of strata control problems was faced.

In one panel where H/t ratio was less than 10, extraction by Russian KM-130 Set (Support density 55 T/m²) was started. When the thickness of hard cover was reduced to 26 m with H/t ratio around 7, first strata control problem was encountered. There was excessive pressure on the chocks and about 40 chocks in the mid-zone were thrown and leaned towards the face and almost collapsed. This phenomenon of collapse of chocks occurred 3/4 times in the panel, whenever the pressure on the chocks was excessive due to low thickness of hard cover. Due to this problem, it was considered that other panels having H/t ratio less than 10 should not be worked with the support (KM-130) and higher support resistance with effective mean load density of at least 80 Tonnes per sq.m should be provided. Higher capacity powered supports (4 X 550 Tonnes) were used which provided higher support resistance of 88 Tonnes per sq.m.

The observations in different panels indicated that roof overlying R-VII Seam at Jhanjra was moderately cavable and intense weighting was not observed. The ratio of maximum mean load density/rated mean load density (MMLD/RMLD) was around 0.8 and convergence was around (7-8)% of seam thickness during first weight and periodic weight. This indicated that support with rated MLD of 55 Tonnes per sq.m. was adequate for the long wall face at shallow cover with H/t ratio more than 10. The ratio of MMLD/RMLD during periodic weighting in case of higher capacity support (4 X 550 T with RMLD as 88 T/m^2) was 0.6. Even when the face was negotiating lower thickness of hard cover and high thickness of alluvium, convergence recorded in such situation was sometimes around (12 to 14) % of seam thickness. The strata control observations data indicated that design capacity of 4 X 550 Tonnes support was more than adequate for adverse condition with H/t ratio around 10.

The percentage of subsidence in panels with lower thickness of hard cover and with higher thickness in weathered rock or/and alluvium was observed to be (57 to 58). This was because of the discontinuous subsidence trough at shallow depth cover where the cracked zone reached the alluvium or weathered rock zone disturbing the subsidence trough.

Table No 1: Minimum thickness of hard cover and weathered mass in panels worked at Jhanjra under Raniganj Coal Field (RCF).

Sl No	Panel Name	Avg. Depth (m)	Min depth (m)	Avg. height of extraction (m)	Min thickness of hard cover (m)	Ratio of minimum thickness of hard cover & height of extraction	Thickness of Weathered material on top (m)
1	W-1	55.0	46	3.4	34.50	10.15	12
2	W-2	55.5	48	3.7	34.60	9.35	14
3	W-3	48.5	45.0	3.5	25.86	7.39	18.93
4	E-1	56.0	50.0	3.2	24.38	7.62	24.38
5	E-2	46.5	40.0	3.2	31.00	9.69	16
6	W-4	42	40	2.4-3.2	16.00	7 app	27

Table No 2: Rock properties overlying E3 panel of Jhanjra Long wall face (Borehole E-3/1)

SI No	Rock type	Run(m)	Bulk density (g/cm3)	Compressive strength (Kg/cm2)	Tensile strength (Kg/cm2)
1	Sandy shale	5.0-8.0	1.90	54.15	-
2	Ferruginous Fg. Sst.	11.34-11.88	2.43	340.79	43.5
3	Shaley Sst.	15.19-15.46	-	-	20.53
4	Mg. Sst.	15.68	2.16	150.03	8
5	Shaley Sst.	17.62	2.07	175.75	4
6	Shaley Sst.	18.55-19	2.28	291.45	-
7	Mg. Sst.	19.2-21.4	2.22	197.18	-
8	Fg. Sst.	21.4-22.0	2.48	334.35	-
9	Cg. Sst.	22.0-22.57	2.11	203.61	-
10	Fg. Sst.	25.12-26	2.46	542.96	106.7
11	Shaley Sst.	26.92-29.5	2.23	231.47	-
12	Fg. Sst.	29.59-30.5	2.51	764.44	112.4
13	Fg. Sst.	30.5-31.12	2.44	251.19	-
14	Shaley Sst.	31.12-32.24	2.20	251.19	-
15	Shaley Sst.	32.43-32.59	-	-	-
16	I.C.	34.26-34.5	2.24	255.05	-
17	I.C.	35.5-35.76	2.18	171	34.5-5.75
18	I.C.	36.07-37	2.08	257	
19	Coal	37.0-38.5	1.36	141.45	-

G - Fine Grained: **MG - Medium Grained:**   **1C – Intercalation**

Table No. 3:

Locate	Thickness of seam	Depth	Avg. Comp. Strength of overlying roof rocks Kg/cm²	Av. RQD of overlying roof rocks	Avg. length of core in cm.	Cavability Index
Jhanjra under RCF	R- VII 2.6-3.0	40-60	386	69	12.7	2426
Jhanjra under RCF	R-VII 2.6-3.0	40-60	277		8.3	3076
Narsamunda	2.8-3.2	72	425	80	10	14116
Bijuri under Son Satpura	Bijuri A seam		168	92		
Jhanjra under RCF	R-VII A	-	254	95		
Jharkhand under Son Satpura	B Seam	-	69	03		

4.0 LONGWALL MINING, PRESENT EXPERIENCE IN SECL

The case studies of 2 projects with 2 panels at Balrampur and 1 panel at Rajendra mine have been presented here for review of performance of Longwall operations at shallow depth of cover. The matter has been presented under following major heads.

    4.1 Geo-technical Details & assessment of support capacity.
    4.2 Case study -1: Panel P-1 & P-2 of Balrampur u/g mine.
    4.3 Case study -2: Panel P-16 of Rajendra u/g mine.

4.1 GEO-TECHNICAL DETAILS & ASSESMENT OF SUPPORT CAPACITY:

4.1.1 Litho log information:

The litho logs of bore holes in Panel P-1, P-2 Of Balarampur (BRP) mine & Panel P-16 of Rajendra (RP) mine are presented in Figures Fig (BRP) 1, Fig (BRP) 2 & Fig(RP)3 respectively.

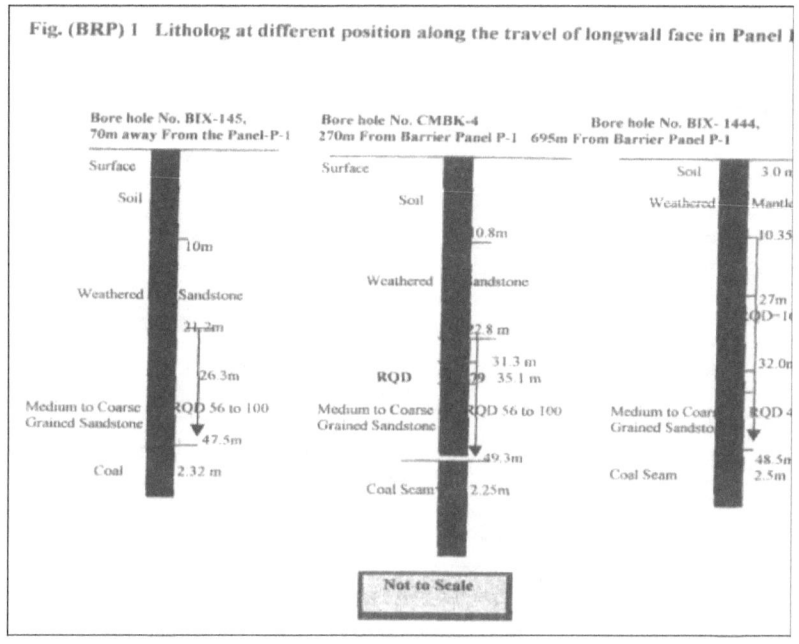

Fig. (BRP) 1 Litholog at different position along the travel of longwall face in Panel

Bore hole No. BIX-145,
70m away From the Panel-P-1

Bore hole No. CMBK-4
270m From Barrier Panel P-1

Bore hole No. BIX- 1444,
695m From Barrier Panel P-1

Not to Scale

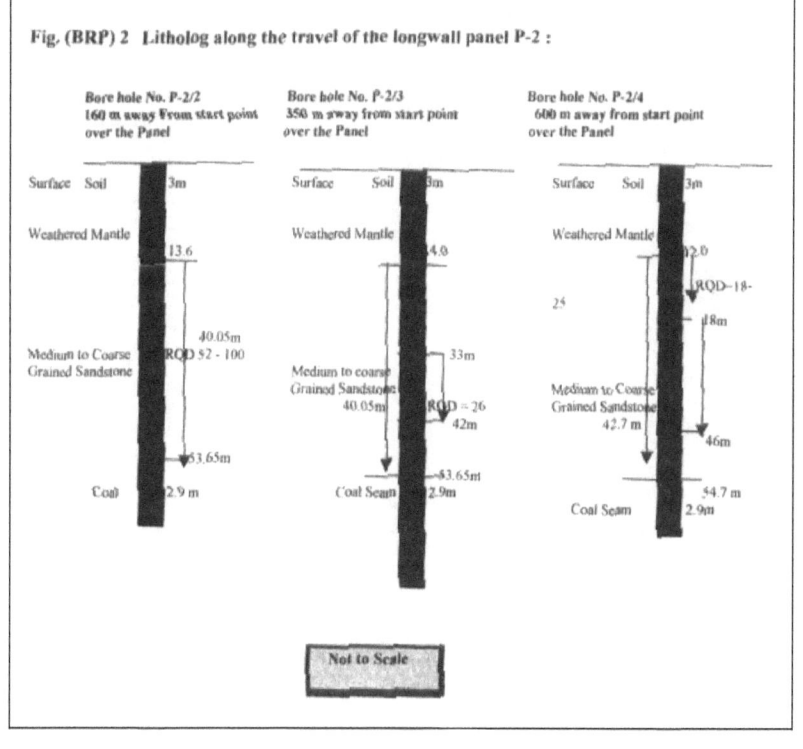

Fig. (BRP) 2 Litholog along the travel of the longwall panel P-2 :

Bore hole No. P-2/2
160 m away From start point
over the Panel

Bore hole No. P-2/3
350 m away from start point
over the Panel

Bore hole No. P-2/4
600 m away from start point
over the Panel

Not to Scale

Fig (RP) 3   **Litholog along the Panel 16 of Rajendra Underground Mine:**

Depth					
0.0	to	3.0	m		Sandy Soil
3.0	to	27.60	m		Coarse Grained Sand stone
27.60	to	28.40	m		Shale
28.40	to	33.0	m		Fine Gr. Sand stone
33.0	to	36.0	m		Coarse Grained Sand stone
36.0	to	39.0	m		Very Coarse Grained Sand stone
39.0	to	41.10	m		Mixture of Med.to Coars.Gr.S.Stone,Coal & shale
41.10	to	42.0	m		Shale
42.0	to	45.0	m		Med. Grained Sand stone
45.0	to	48.0	m		Coarse Grained Sand stone
48.0	to	51.0	m		Med. to Coars.Gr.S.Stone
51.0	to	54.0	m		Coarse Gr.S.Stone
54.0	to	57.0	m		Coarse. Gr. to very OCRs. Gr. sand stone
57.0	to	60.0	m		Coarse Gr. S.Stone
60.0	to	63.0	m		Coarse. Gr. to very Coarse. Gr. Sand stone
63.0	to	66.33	m		Coal, Burhar VI Bottom seam, with the proposal
66.33	to	70.50	m		Med. to Coarse.Gr.S.Stone

The summarised litholog information of Panel P-1 & P-2 of Balrampur and panel P-16 of Rajendra mine are given in table(BRP)4 and table(RP)5 respectively.

Table No. (BRP) 4 :   **BALRAMPUR NO. 10 & 12 INCLINES**

	Panel P-1		Panel P-2	
	minimum	maximum	minimum	maximum
Soil/ weathered sand stone in meters	21.2	22.8	13.6	14.0
Depth of cover in meter	47.5	49.3	53.6	54.0
Medium to coarse gr. Sst (hard cover) h (in meter)	26.3	26.5	40.05	40.05
Seam thick in m (extracted thickness in m(h))	2.4 (2.25)	2.4 (2.25)	2.25 (2.25)	2.25 (2.25)
h/t - Hard cover/ seam thickness ratio	10.9	11.0	17.8	17.8

Table No (RP) 5: RAJENDRA UNDERGROUND MINE

	BH-1790 (30m)	BH-1734 [105 m]	BH-1792 (180 m)	BH -1793 (300 m)	BH-1794 (400 m)	BH - 3 (600 m)	BH-1733 (1090 m)
All & weathered Rock (in meter)	15.25	15.25	15.25	12.20	12.2	21	
Cover/Depth (in meter)	48 80	51.5	51.25	46.80	46.63	47	
Hard St. Cover (h) (in meter)	33.55	36.25	36.0	34.60	34.43	36	
Ht. of extraction (t) (in m`eter)	2.55	2.55	2.55	2.55	2.55	2.55	
Ratio h/t	13.2	14.2	14.1	13.6	13.5	14.1	

4.1.2: PHYSICO-MECHANICAL PROPERTIES OF STRATA IN DIFFERENT PANELS AT BALRAMPUR MINE:

Table No. (BRP) 6:

Depth(m)	R Q D of the beds		Comp. Strength in MPa		Tensile Strength in MPa	
	P-l	P-2	Panel P-1	Panel P-2	Panel P-1	Panel P-2
13.50-15.60	00	73	0	0	0	2.1
15.6-18.65	00	81	3.28	10.4	0.57	3.7
18.65-21.00	00	96	5.5	13.4	0.5	1.9
21.00-24.75	00	91	7.7	25.3	1.04	2.6
24.75-27.25	57	98	7.4	11.4	0.94	2.3
27.25-30.30	81	95	5.24	15.8	0.76	2.1
30.30-33.35	71	95	4.9	34.9	0.73	5.4
33.35-36.40	92	100	5.4	16.5	0.5	1.3
36.40-39.45	85	74	8.46	31.4	1.18	2
39.45-42.50	96	63	12.11	33.1	- 1.46	2.7

42.50-45.55	100	100	12.8	19.6	1	2.7
45.55-47.60	69	100	6.92	24.8	0.35	1.2
47.60-50.65	30(Coal)	100	28.35(Coal)	12.2	2.11 (coal)	3.9
50.65-53.65		100	-	12.2	-	6.4
53.65-56.55	-	50(coal)	-	27.9(coal)	-	3.6 (coal)

PANEL P-1

The average compressive strength was 7.2 MPa (73.9 kg/cm^2) and the maximum compressive strength of 12.8 MPa (i.e. 130.6 Kg/cm^2) was at (42.5 to 45.5) m (about 3 m thick). The average tensile strength was 0.82 MPa (8.4 kg/cm^2) with maximum at 1.46 MPa (14.9 kg/cm^2) at 39.45 to 42.5 (about 3 m) but not coinciding with maximum compressive strength.

At 160 m travel, the strata lying between depth of 27.25m & 45.55m was of high RQD, average being 88 % (range 71 to 100). This 18m high RQD stratum was 2m above the roof level. At 270 m travel, high RQD - 75 % (range 65 to 85) was above 5 m of roof. At 695 m travel, high RQD - 96 % for about 6 m, about 4m above the roof was observed.

Higher average RQD (81-90)%, lower average compressive strength (73 to 205) kg/cm^2, lower average tensile strength (8.4 kg/cm^2) and higher h/t ratio compared to strata met at Jhanjra mine under RCF were observed.

PANEL P -2

In Panel P-2 the average compressive strength was 20.1 MPa (204.8 kg/cm^2) and the maximum compressive strength was 43.66 MPa (355.9 kg/cm^2) at depth of 30 m to 33 m (about 3 m thick). The average tensile strength was 2.95 MPa (30.14 kg/cm^2) with maximum 5.11 MPa (65.2 kg/cm^2) at 50.6 m to 53.6 m (about 3 m thick) just above the coal seam.

At 160 m, high RQD - 90 % (range 63 to 100) occurred for about 40m just above the roof. At 350 m, high RQD - 94 % for about 12 m, about 14 m above the roof and at 600 m, high RQD - 71 % for about 7 m, about 9 m above the roof were observed. Compared to the strata in Panel P-l, the strata in Panel P-2 had higher compressive strength, higher tensile strength, lower RQD and higher h/t ratio.

PHYSICO-MECHANICAL PROPERTIES OF ROCK IN PSLW PANEL 'P - 16' OF RAJENDRA UNDERGROUND MINE.

Table No. (RP) 7: *BH - 1734 (105 M)*

DEPTH (M.)	LITHOLOGY	R.Q.D.	UNIAXIAL COMPRESSIVE STRENGTH (MPa)	TENSILE STRENGTH (MPa)
15.25-18.3	FG-MGSST	60	14.9	1.4
18.30 - 21.35	MG-CGSST	28	7.0	1.0
21.35-24.4	CG-VCGSST	59	5.4	0.7
24.40 - 25.95	VCG-SST 0.89-COAL	65	10.10	1.5
25.85-29	0.80-SHALE FG-CGSST	50		
29.00 - 32.05	CG-VCGSST	28	6.9	0.99
32.05-35.1	CG-SST	43	7.9	0.92
35.10 - 38 15	-	29	8.4	0.7
38.15-41.2	MG-CGSST	49	8.4	0.7
41.20-44.25	CGSST	50	6.9	0.5
44.25 - 47.3	VCG-CGSST	56	5.9	0.6
47.30 - 50.35	VCG-SST	51	5.9	0.6
50.35 -53.4	CG-SST 1.90 COAL	23		
53.40 - 56.	1.75 COAL MGSST	36	13.7	

CONNOTATIONS:

F.G.S.S.T.	-	*FINE GRAINED SANDSTONE*
M.G.S.S.T.	-	*MEDIUM GRAINED SANDSTONE*
C.G.S.S.T.	-	*COARSE GRAINED SANDSTONE*
V.CG.S.S.T.	-	*VERY COARSE GRAINED SANDSTONE*

PANEL: P-16

The average compressive strength of overlying strata was 81.3 kg/cm^2 with max 152 kg/cm^2 at (15-18) m from surface which was less compared to Balrampur mine.

At 150 m travel, average compressive strength was 98.7 kg/cm^2 with max 260.1 kg/cm$^{2:}$ at 41 to 42 m from surface. At 600 m travel, average compressive strength 119.9 kg/cm$^{2:}$ with max 223 8 kg/cm$^{2:}$ at 31 to 32 m from surface were observed.

Maximum RQD was 70 at 30 to 33 m from the surface and in the rest it was less than 70. *At 150m travel,* Max RQD 84 for 0.9 m strata followed by 71 for 3 m strata & at 600 m travel, max RQD was 65 for 3 m strata.

The RQD was comparatively less than strata met at Balrampur mine. However, the high RQD was near the coal seam. The strata was moderately cavable (Caving index - 1118 - 3513) with maximum value between 2000-4000. It was watery and there were weak planes in sand stone bed. The ratio of hard cover to extraction thickness (H/t) was about 14.

4.1.3 ASSESSMENT OF SUPPORT CAPACITY:

(1)   *Thickness of Extraction:* 2.5 meter.

(2)   (a) As per CMRI study, the effective support density assessed based on geo-technical data of bore holes outside 1st panel was:

     (i)   **(After cut)- 51 t/m^2**
     (ii) **(Before cut)-59 t/m^2**

(b) Recommended Rated load density, considering 90% hydraulic efficiency of the system:

     (i)   **(After cut)- 57 t/m^i**
     (ii) **(Before cut)-66 t/m^2**

(c) Rated load density of the supports purchased By SECL was:

     (i)   **(After cut) 69.2 t/m^2**
     (ii) **(Before cut) 80.3 t/m^2**

(d) Rated load density of the upgraded supports
(After increasing the yield setting):

  (i)   **(After cut)** -   **79 t/m²**
  (ii) **(Before cut)** -   **91.6t/m²**

(e) Rated load density of the supports after modification/up gradation to 600 tonnes
Capacity was:

  (i)   **(After cut)-92.3 t/m²**
  (ii) **(Before cut)-107 t/m²**

(f) Ratio of e (rated load density) / b (recommended) -**1.62**

In all the long wall panels under review the support used had the following specifications

(a) Original support capacity: ....   **50 Tonnes**

(b) Support capacity of the up rated support after
Changing the yield pressure setting.   **513 Tonnes**

(c) Rated support resistance originally provided
  (i) **(After cut) - 69.2 t/m²**
  (ii) **(Before cut) - 80.3 t/m²**

(d) Rated support resistance provided after   **(After cut) - 79 t/m²**
changing the yield pressure settings:......   **(Before cut) - 91.6 t/m²**

4.2. CASE STUDY -1: Panel P-1 & P-2 of Balrampur U/G mine

The details of panel geometry for panels P-1 & P-2 are given in Table no. (BRP) 8:

Parameter	Values		Remarks
	Panel P-1 (Extracted)	Panel-2 (Extracted)	
Panel length	736 m	857	
Width	150 m	150	
Cover	47-51m	52 to 54	
Seam thickness	2 to 2.8 m	2 to 2.8 m	
Thickness of extraction	2.4m to 2.25m	2.25	
Gradient	1 in 51 N 30^0 35'E	1 in 51 N 30^0 35'E	
Immediate roof	0.2 to 0.4 m shale followed by coarse grained sandstone	0.2 to 0.4 m shale followed by coarse grained sandstone	
Chain pillar	55m to 70 m ×20m (centre to centre)	55m to 70 m ×20 m(centre to centre)	
Degree of gassiness	I degree	I degree	

The technical features of supports & shearers are given in table No.(BRP) 9&10.

Table No. (BRP) 9: TECHNICAL DETAILS OF SUPPORTS:

SI. No.	Parameters	Values
1.	Type of the support	4 legged chock Shield, ZZ 4400/14/27(Chinese)
2.	Height of the support	1.4 to 2.7 m
3.	Width of the support	1.42 to 1.59 m
4.	Central distance	1.5 m
5.	Setting pressure	29 MPa
6.	Yield pressure	40 MPa

7.	Maximum load bearing capacity	513 Tonnes
8.	Maximum Hydraulic travel	671 mm.
9.	Provided with positive set valve	29 MPa

Table No. (BRP) 10: TECHNICAL DETAILS OF SHEARER:

a) General description	Values
Machine height	1150 mm
Center distance of ranging arm	5070 mm
Center distance of drum	8432 mm
Cut depth	630 mm
Drum diameter (mm)	1600
Maximum mining height (mm)	3000
Under cutting (mm)	400
Best range of mining height (mm)	1.8-2.7
b) Electric Motor	Values
Model	YBCS-375
Power	375 kW
Voltage	1100 V
c) Haulage system	Values
Type	Pin rack chainless haulage
Haulage speed	0.61 m/min
Haulage force	450 KN
d) Cutting system	Values
The length of ranging arm	1681 mm
Swivel angle of ranging arm	17°-150°

Double Ended Ranging Drum Shearer of drum diameter 1.6 m and web 0.6 m having cutting range up to 3 m was installed at the face. In the beginning, setting pressure of the powered support was kept at 30 MPa while bleed pressure at 35 MPa. The maximum operating pressure of the power pack was 31.5 MPa, whereas it was being operated at 30MPa. Four legged Chock Shields 103 in numbers had been installed at the face in Panel P-1. After the incidence of failure at 160m advance position of the face, the setting of the bleed valves was changed to 40 MPa, thereby increasing the support capacity to 513 Tonnes from 450 Tonnes. When the bleed valve setting was raised from 35 to 40 MPa, the support resistance offered at the Long wall face also increased from 80 Tonnes per m^2 to 92 Tonnes per m^2 (before cut) & 69 Tonnes per m^2 to 79 Tonnes per m^2 (after cut).

The management proposed to continue with the same chock shields supports up-rated to 513 Tonnes by increasing the setting of the bleed valves from 35 MPa to 40 MPa in other projects.

4.2.1 PANEL: P-1

4.2.1.1. CHRONOLOGICAL EVENTS:

The panel was started on 11-5-98 and the first main fall occurred when the face advanced 80m. with an exposed area of about 12000sq.m., 16 supports in the mid zone collared. After recovery of the supports, face operations continued. The second main fall and weighting occurred when the face advanced to 160m. with an area of exposure 24000 sq.m. A number of powered supports started bleeding and maximum cumulative convergence of one of the supports was found to be 630mm. The peak leg pressure in some of the supports located in the central zone showed above 40 MPa resulting in collaring of 13 supports. Crack was observed in the roof in between the face and tip of the front canopy. On scrutiny of the yield valves of the supports it was observed that a number of yield valves failed. Face had to be kept stopped for 46 days. A detailed examination of individual supports revealed the failure of the following components.

(i)   13 supports collared & canopy got damaged
(ii) Buldging of sector rings of powered support leg.

(iii) Bending of inner tubes of legs

(iv) Leakage in non-returns valve.

(v) Mal-functioning pressure gauges

Careful scrutiny of the failed components revealed some defects and these components were replaced.

The incident was critically examined by the experts of Directorate General of Mines Safety (DGMS), South Eastern Coal Fields Limited (SECL), Central Mining Research Institute (CMRI), Central Mine Planning and Design Institute (CMPDIL) and China National Coal Mining Engineering Equipment (Group) Corpn.(CME) and the following decisions were taken.

(i)   To increase the capacity of the supports by increasing the yield pressure from 35 MPa to 40 MPa.

(ii) To adjust the mechanical extension of the supports in order to provide maximum hydraulic travel.

(iii) To install positive set valves in the leg circuits to ensure desired initial setting load.

(iv) To introduce induced caving by deep hole blasting from surface in order to reduce overhang & excessive loading on support. Initially the objective should be not to exceed overhang beyond 30 m.

The face was again started after implementation of most of the above decisions. Deep hole blasting from surface was started at 178m travel to induce caving when the face was at 191 m travel. Blasting interval was maintained at 15 m from 240 m to 390m. Subsequently the face continued up to 451m. Blasting interval increased to 30m from 451m up to 589m. Major weighting occurred at 515m travel when cumulative convergence of 410 mm was recorded. Again after 589 m blasting interval was reduced to 15 m and continued up to 679 m travel.

After the face had reached 717 travel, preparation for salvaging was started. The decision was taken to support the immediate roof by full column grouted bolts set at 1 m interval. The final position of the long wall face was decided to be at 725 m, but after the face had reached 720 m another periodic weighting occurred and the face had to be advanced further up to a stable ground. The face ultimately was stopped at 736 m.

During deep hole blasting from surface, the leg pressure measurements and peak particle velocity were recorded in underground regularly. The peak particle velocity (PPV) at a distance of 15 m towards the face on surface was 67 mm/sec. The PPV at the centre of the face was more than 149 mm/sec where as it was 51 mm/sec in the main gate and 31 mm/sec in the tail gate. The rear edge of the chock shield was kept at a radial distance of about 22 m and at a horizontal distance of 11 m from the explosive column.

The average production was about 2650 Tones per day with maximum up to 4600 Tones in a day. The maximum face advance achieved per day was 9 m.

4.2.1.2 LOAD ON SUPPORTS:

At the beginning, the support capacity was 450 Tones with support resistance of 69 t/m^2 (after cut). After the major weighting at 160m the support capacity was up rated to 513 Tones (with support resistance of 79 t/m^2 after cut). During the first major weight at 80m of face advance 16 supports in the mid zone collared. When the face was at 160m the second major weight was experienced & 13 supports were collared and damaged. The average pressures in the overall face are shown in figures (BRP)- 4 & (BRP)- 5. After the first major weighting, the MMLD reached yield load during eleven periodic weightings. After the supports were up rated to 513 Tones capacity (support resistance 79 t/m^2 after cut) and introduction of blasting from surface, loading on supports increased. During subsequent operation in the panel the supports were not damaged due to sudden loading.

4.2.1.3 INDUCED CAVING BY BLASTING FROM SURFACE:

Induced blasting from the surface was resorted to at 178 m from the start of the face when the face position was at 191 m travel. Regular blasting was conducted at an interval of 15m. Borehole details showed the presence of hard massive strata with RQD in the range of 69 to 96 at a depth of 30 m to 40 m from surface. 100 mm dia. blast-holes were drilled up to a depth of 35 m in general and one up to a depth of 40 m. Depending on the location and massiveness of the strata, nos. of decks were decided. While length of the bottom deck was kept 3 m with inter-deck stemming length not less than the length of the charge column, the length of the upper deck was decided as 2.5m.

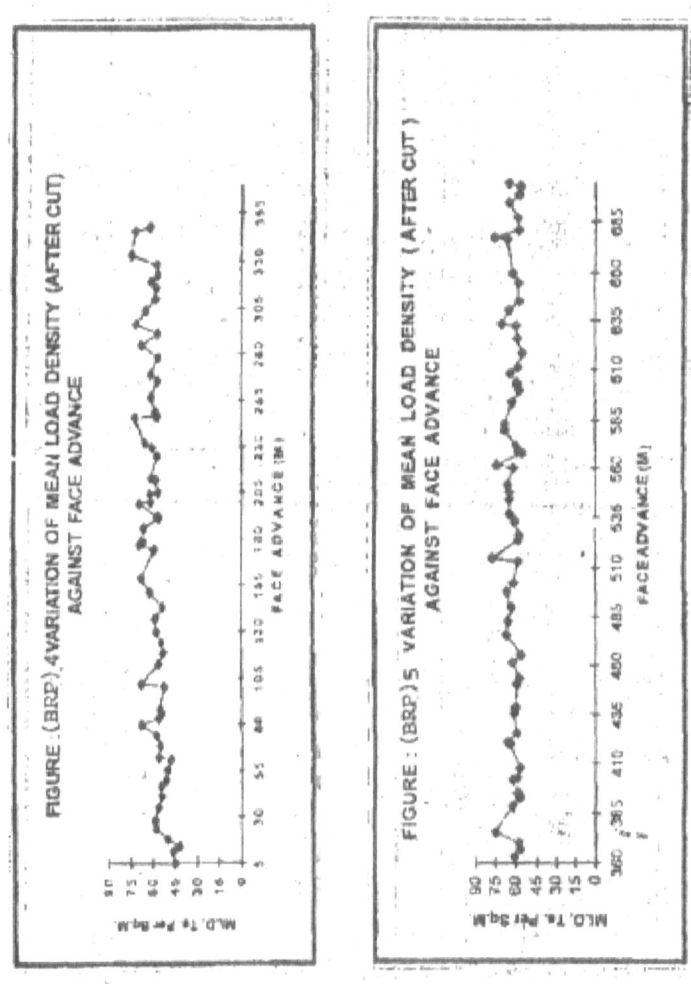

The interval between the blasting rows was maintained at 15 m but from 463 m to 602 m the blasting interval was made to vary from 18 m to 32 m to see the effect of increased row interval. The number of holes in a row were (9-13) in general.

Details of the explosives charge column are given in table No (BRP) -11.

TABLE NO. (BRP) 11:

SL No.	Description	Specification
1	Type of explosive	Acquadyne
2	Cartridge dia	83 mm
3	Cartridge weight	2.78 Kg.
4	Blast hole diameter	100 mm
5	Loading density	9 Kg/m
6	Detonation velocity	3400 - 4300 m/sec
7	Density	1.12 to 1.2 mgs/cc
BOTTOM DECK		
1	Length of the charge column	3 m
2	Nos. of cartridge	10
3	Total weight of explosives	27.8 Kg.
TOP DECK		
1	Length of column	2.5 m
2	Nos. of cartridges	8
3	Total weight of explosives	22.2 Kg.
4	Total no. of hole per blasting	9 to 13 nos.
5	Total explosives charge per hole	50 Kg. (Approx.)
6	Length of span of blasting	60 to 70 m

The impact of induced caving on the following parameters has been discussed below in details:

    (a) **Loading on supports**
    (b) **Periodicity of weighting.**
    (c) **Convergence.**
    (d) **Subsidence.**

(a) IMPACT OF INDUCED CAVING ON LOADING:

The trend of loading on supports before blasting and after blasting is shown in Table NO.(BRP)12.

TABLE NO. (BRP)- 12:

Load in T/m^2		45-65	65-70	70-75	75- YL	Total
Overall	Frequency	98	13	8	5	124
	Percentage	80	10	6	4	100
Before Blasting	Frequency	29	2	-	-	31
	Percentage	94	6	-	-	100
After Blasting	Frequency	69	13	6	5	93
	Percentage	74	14	7	5	100

Before blasting, loading was confined between (45 – 65) Tones/m^2 in about 94 % cases and in rest 6% cases loading was more than 65 T/m^2. When blasting was resorted to and the support resistance increased to 79 T/m^2, 74 % loading was up to 65 T/m^2, 14 % between 65 to 70 T/m^2 and 7 % loading was between (70-75) T/m^2 and 5 % loading was between (75-79) T/m^2. It could be concluded from the records that blasting had resulted in increase of MLD beyond 65 t/m^2 and in one occasion the MMLD reached the RMLD. In consideration of over all i.e. both before & after blasting, then 80 % loading was up to 65 T/m^2 and rest 20% loading was more than 65 T/m^2.

The ratio of hard cover and thickness of extraction (H/t) was about 11 and thickness of weathered sand stone and soil cap on top was about 21 m. These conditions might have resulted in formation of discontinuous subsidence trough and resulted in increased loading on supports when surface blasting was resorted to.

(b) IMPACT OF INDUCED CAVING ON PERIODICITY OF WEIGHTING:

TABLE (BRP)- 13: INTERVAL BETWEEN THE PERIODIC WEIGHTINGS:

		(5-l0)m	(10-20)m	(20-30)m	> 30 m	Total	
Before Blasting	Frequency	2	3	3	1	9	Wide range.
	Percentage	23	33	33	11	100	
After Blasting	Frequency	2	15	9	1	27	Concen trated Range.
	Percentage	7	56	33	4	100	
Complete Panel.	Frequency	4	18	12	2	36	
	Percentage	11	50	33	6	100	

The Table No. (BRP)- 13 above show the periodicity of weighting under different conditions i.e. before blasting and after blasting and the overall condition with both the situations. It was observed that before adoption of blasting about 56 % occurrence of weighting was between (5-20) m and 33 % between (20-30)m and 11 % beyond 30 m where as the figures were 63 % between (5-20)m, 33 % between (20-30)m and only 4 % beyond 30 m when blasting was adopted

It could be concluded from the records that blasting resulted in reduction of periodicity of weighting beyond 30 m and increase of regularity in weighting between (10-20) m.

(c) IMPACT OF INDUCED CAVING ON CONVERGENCE:

The figure (BRP)- 6 shows the maximum cumulative convergence during periodic weightings. It was observed that after resorting to blasting the cumulative convergence in general was reduced except during one major weighting when the face was at 515m cumulative convergence increased to 410 mm. During this period with face at 515m blasting interval was increased from 15 m to 30 m and MMLD also reached the Rated Mean Load Density (RMLD).

The figure (BRP) 7 shows the convergence along the face at 515 m during major weighing. The convergence during cutting is separately shown and the cumulative convergence curve is also drawn. The cumulative convergence curve shows that the average cumulative convergence at the mid zone was higher than the convergence on both ends.

When the blasting interval was 15 m, cumulative convergence was in the range of 25 mm to 108 mm. But when the interval was raised to 30 m.cumulative convergence increased to the range of 40 mm - 410 mm.

(d) INFLUENCE OF INDUCED CAVING ON SUBSIDENCE:

Figure (BRP) 8 shows the movement of a surface subsidence stations (point) before blasting and after blasting & Figure (BRP) 9 shows the subsidence profile across the panel. It may be concluded that

(i) The magnitude of negative traveling angle of draw was reduced i.e. the subsidence point advanced towards the face after blasting was resorted to.

(ii) The blasting in some cases had limited impact in vertical movement of stations and smoothening of subsidence profile.

(iii) The development of subsidence profile for extraction in shallow depth of cover gave indication of approximate extent of overhang of strata over goaf.

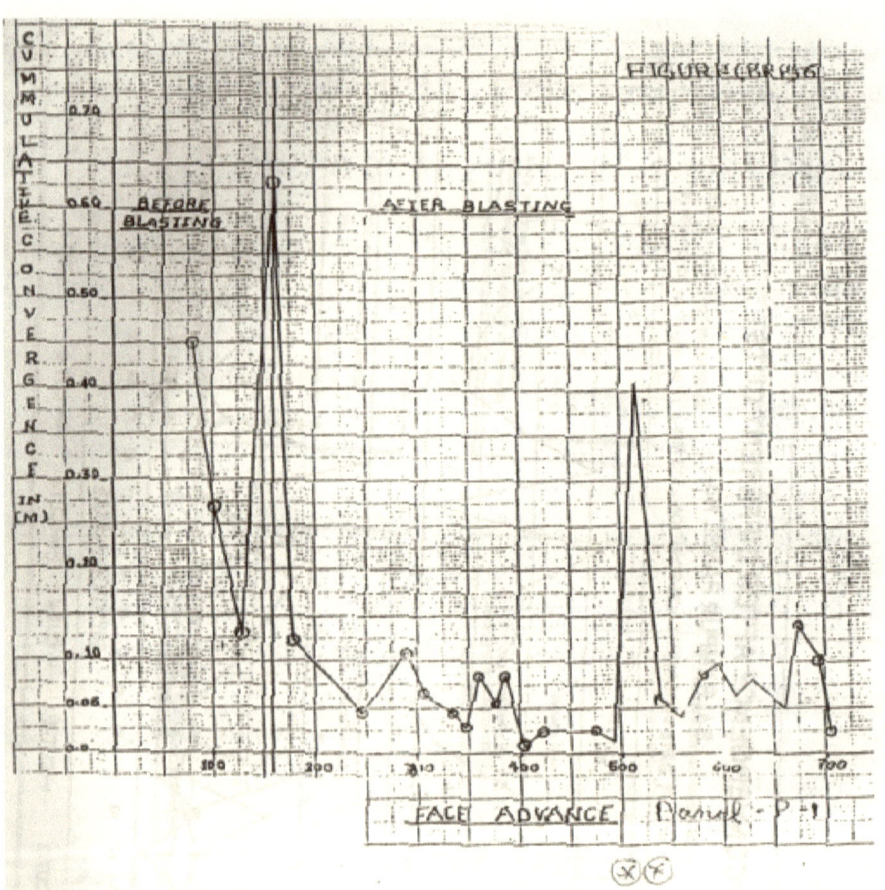

FIGURE (BR) 6

CUMMULATIVE CONVERGENCE IN (M)

BEFORE BLASTING

AFTER BLASTING

FACE ADVANCE (Panel - P -1)

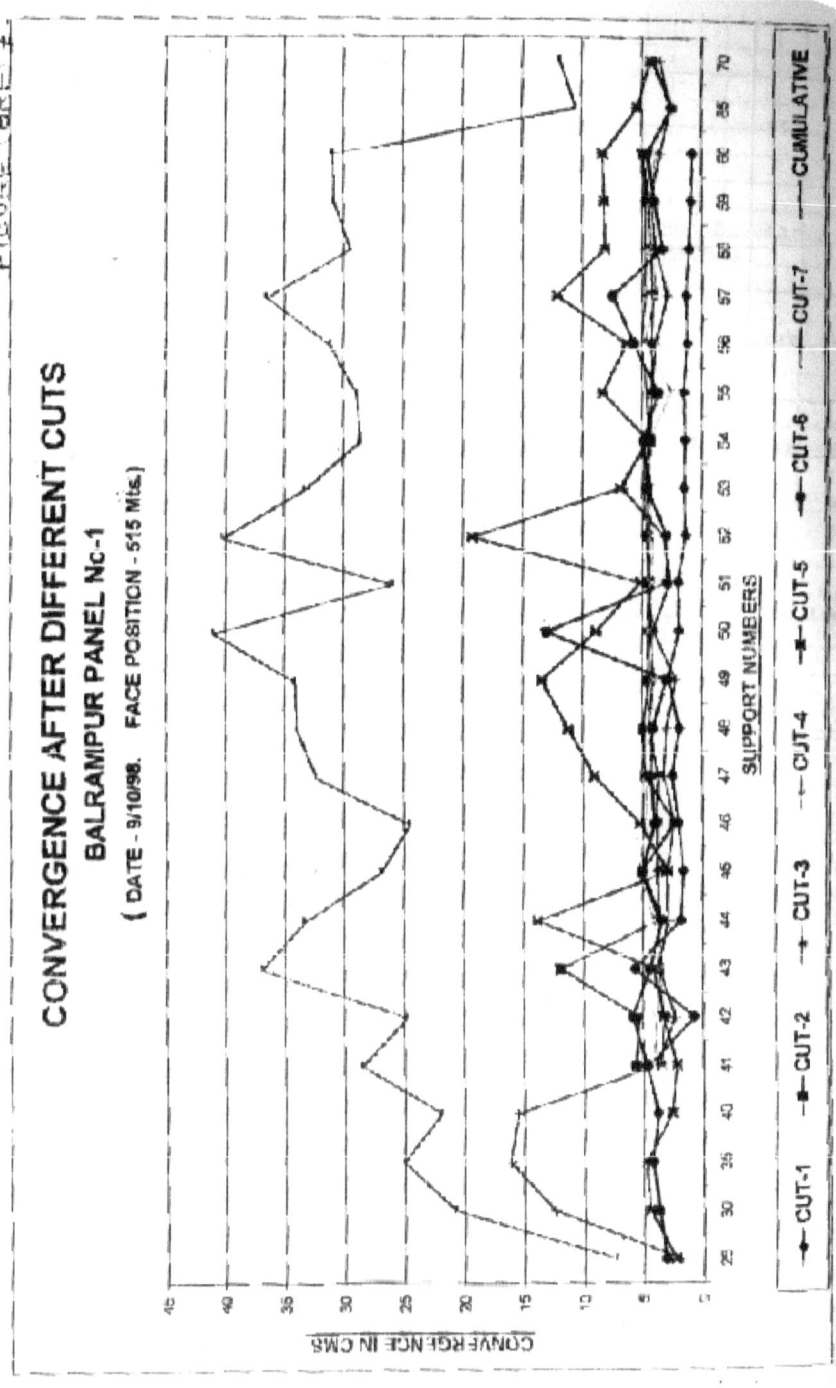

FIGURE : 1

CONVERGENCE AFTER DIFFERENT CUTS

BALRAMPUR PANEL Nc-1

(DATE - 9/10/98.   FACE POSITION - 515 Mts.)

CONVERGENCE IN CMS

SUPPORT NUMBERS

CUT-1   CUT-2   CUT-3   CUT-4   CUT-5   CUT-6   CUT-7   CUMULATIVE

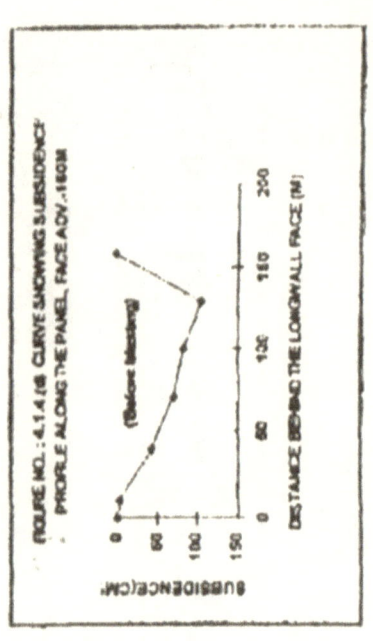

FIGURE NO. : 4.1.4.(c) CURVE SHOWING SUBSIDENCE
PROFILE ALONG THE PANEL , FACE ADV - 102M

(Before blasting)

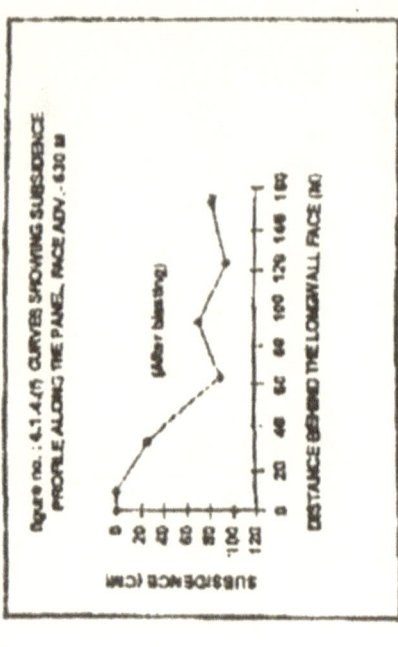

FIGURE NO. : 4.1.4.(d) CURVE SHOWING SUBSIDENCE
PROFILE ALONG THE PANEL, FACE ADV -160M

(Before blasting)

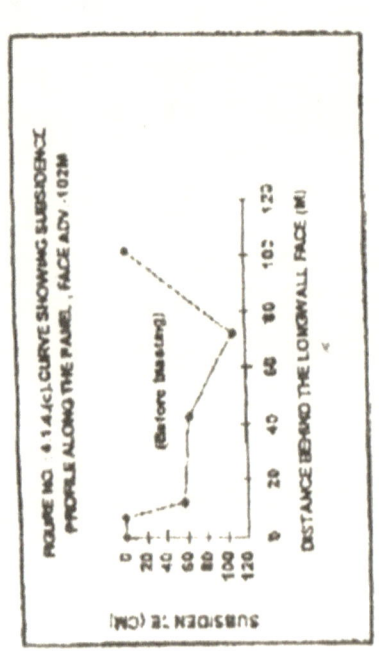

FIGURE NO : 4.1.4.(e). CURVE SHOWING SUBSIDENCE
PROFILE ALONG THE PANEL, FACE ADV - 446 M

(After blasting)

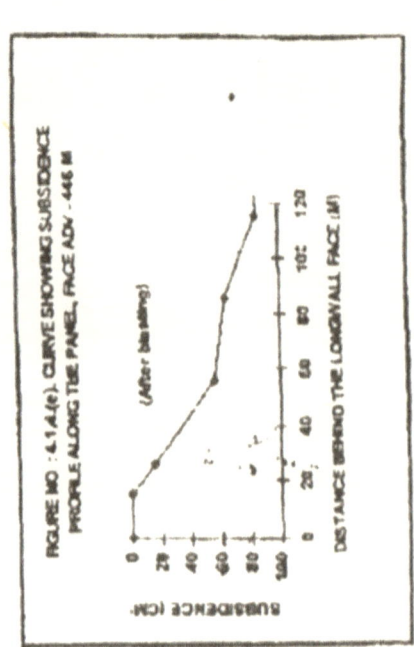

figure no. : 4.1.4.(f) CURVES SHOWING SUBSIDENCE
PROFILE ALONG THE PANEL, FACE ADV - 630 M

(After blasting)

FIG(BRP)-8

FIGURE NO.: 4.11(a): SUBSIDENCE PROFILE ACROSS THE PANEL
ALONG 230 M LINE FROM START OF THE PANEL

4.2.1.4 CONVERGENCE AT THE LONG WALL FACE:

TABLE NO. (BRP) 14:

Face position in meter	Max cumm. Conv. mm	Avg. cumm. Conv. mm/mt	Max. Conv. In one cycle (mm/mtr)	Remarks
80*	450			supports collared and damaged
160*	630			
244	44	25	52	
515	410	98	320	major weighting
659	41	17	50	

The above Table shows the convergence during individual cut, cummulative convergence in mm and cummulative convergence per meter of face advance. The cummulative convergence of 98 mm/mt face advance and maximum rate during one cut of 320 mm/mtr were recorded when the face was experiencing major weighting at 515 m.

The high convergence rate during particular cut indicate the supports were under pressure and for successful operation the advance of the face was a must. However when the face advance exceeded 9m to 10m per day higher load was experienced on the supports during periodic weighting.

4.2.1.5 INFLUENCE OF STRATA IN CAVING:

From the records of bore hole extensometer, it may be concluded that when the face reached beneath the instrumented bore hole, there was no movement of anchors. When it is crossed by about 10m movement started. When it crossed by 35m, caving up to 16m i.e. 9 times the thickness of extraction was recorded. The movement of anchors and ultimate caving depends on the type of strata. At lower RQD, the strata cave in early. The bulking factor calculated on the basis of observation worked out to be 1:1.

The blasting was conducted at different intervals i.e 15m, 20 m & 30 m but most of the times at 15 m interval. When the RQD of the strata was around 50, the surface cracks used to cross the holes in the central position and the holes were getting damaged; in such condition single deck blasting had to be restored to.

4.2.1.6 SUBSIDENCE AND ITS EFFECT ON CAVING:

The depth of coal seam was 48m and thickness of extraction was 2.25 m. The subsidence profile across the travel of the face at 210 m has been shown in Figure (BRP) 9. The maximum subsidence 118cm. was developed after face advance of 171m. We may say the maximum subsidence of 52.4% developed when the face advance was 3.4D and the length of face was 2.9D. The max. angle of draw on the dip side was +20° 34' and on the rise side +16° 16'. The concept of critical width is probably not applicable at shallow depth of cover. This has also been observed in subsidence observation carried out in mines under Jharia Coal Field(JCF) at shallow depth of cover as reported in published literature (2).

The following observations were made during study of subsidence phenomenon in the panel:

(1) In low RQD regions, subsidence used to reach closer to the Long wall face and the cracks on surface had appeared up to the rear edge level of the canopy. The frequency of periodic weighting had also increased. But the pressure profile of leg circuits did not change except reduction in convergence.

(2) The gradient of the subsidence profile across the Long wall face in the central portion used to become flat at a distance varying from 70 to 120 m behind the Long wall face. In the area of low RQD zones, the subsidence used to be quicker and was maximum at lesser distance behind the Long wall face (56 m).

(3) The subsidence at faster rate following the faster advance not more than 8m/day of the Long wall face had helped reduction in load getting transmitted on to the supports as well as at the long wall face.

4.2.1.7 INSTRUMENTATION IN GATE ROADWAYS AND STRATA CONDITION:

Load cells and convergence recorders were installed in Main and Tail gate and records were maintained. Maximum convergence of 5.2 mm and maximum increase of load up to 3 Tonnes were recorded. As such there was no problem of roof control in gate roadways with 40 Tonnes hydraulic props set on either sides with girders set on the props at 1 m interval.

4.2.1.8 ROUTINE CONDITION MONITORING OF POWERED SUPPORTS:

Table (BRP) 15, below shows the health of the powered support during different periodic weightings. It was observed that during the first and second main weights, the unhealthy leg circuits were more than 20 %. Subsequently number of actions as discussed earlier was initiated and that had resulted in improvement of performance of powered supports.

However, during a major weight when the face was at 698 m the unhealthy leg circuits were again more than 10 % i.e. the maximum permissible level for efficient performance of powered support.

**TABLE NO (BRP) 15. LEG CIRCUIT CONDITION MONITORING-
(TOTAL NUMBER OF LEG CIRCUITS -206)**

Sl No.	DATE	FACE ADVANCE (m)	% OF LEG CIRCUITS IN HEALTHY CONDITION	% OF UNHEALTHY LEG CIRCUITS
1	28/29.5.98	80	79.61	20.39
2	06/07.6.98	102	83.5	16.50
3	17/18.6.98	160	78.16	21.84
4	19/20.7.98	171.9	98.06	1.94
5	26/27.7.98	171.9	98.54	1.46
6	02/03.7.98	175.09	99.51	0.49
7	09/10.8.98	186.3	99.51	0.49
8	15/16.8.98	203.59	99.51	0.49
9	02/03.9.98	316.05	99.51	0.49
10	17/18.9.98	395.00	98.54	1.46
11	01/02.10.98	464.03	93.69	6.31
12	19/20.10.98	568.08	92.72	7.28
13	01/02.11.98	632.35	97.09	2.91
14	08/09.11 98	667.06	97.57	2.43
15	15/16.11.98	698.00	87.38	12.62

4.2.2 PANEL P-2 - BALRAMPUR No. 10/12 Inclines

4.2.2.1 CHRONOLOGICAL EVENTS:

The face length was 150 m and travel was 910 m but extractable travel was 857 m. The panel geometry and the details of face supports and machineries were presented earlier in table no (BRP) 8,9 & 10.

The long wall panel P-2 was commissioned on 4-3-1999. Regular deep hole blasting from the surface were conducted from the beginning of the face operation.

Deep hole blasting from surface at 17m, 32m, 47m with a relief blast at 40m were conducted but there was no subsidence. It was then decided to change the blast design to double row with continuous staggered column at 58m and 62m. When face was at 73m caving occurred immediately after blasting and again it reoccurred after 3m advance of face.

When the face was at 87 m. first main weight was recorded in underground. The second major weight was recorded when the face was at 151 m. Cracks and cavities were observed in roof covering 45 m length of face at the mid zone. Surface blasting was continued at an interval of 15 m from the start of the panel but two rows of blasting was skipped when the face was at 137 m and 152 m and again blasting was resorted to at an interval of 15 m up to 272 m. When the face was between 116 m and 179 m the fall was regular and the periodicity of weighting was reduced to about 8m. Surface blasting continued and the regular periodic weightings were observed.

When the face was advanced to 383 m, major periodic weight was observed, and face had to be kept stopped for 17 hours due to breakdown in A.F.C. Major Part of the cumulative convergence was due to stoppage of the face. Blasting interval between rows was increased to 20 meters between 392 m and 412 m and then again blasting interval was increased to 30 m. The blasting interval was again reduced to 20 meters from 412 m up to 582 m. The major weighting was observed when the face was at 463 m. Blasting interval was reduced to 15 m from 582 to 612 m and then again blasting interval was increased to 20 m.Long wall operations continued with occurrence of periodic weightings and the rate of face advance was regulated to control increase in pressure on supports. Long wall face continued up to 857 m from the start.

The average production achieved was 2882 Tonnes per day with highest production of 4600 Tonnes in a day. The average rate of advance was 6m/day with maximum advance achieved was 9 m/day.

4.2.2.2 LOADING ON SUPPORTS:

The table below shows the loading frequency and range of Mean Load Density (MLD). The MLD did not exceed 75 T/m^2. Seventy percent of loading during periodic weighting was at less than 65 T/m^2 and 30 % of loading was in between (65-75) T/m^2

Table No (BRP) 16: Loading frequency & MLD during periodic weighting:

MLD in T/m²	<65	65-70	70-75	>75 to YL	Total
Frequency	25	8	3		36
Percentage	70	22	8		100

The figure No. (BRP) 10 at face positions of 87M, 101M, 116M, I38M shows that MLD sometimes was more in top zone and sometimes in bottom zone but all the time was less than the middle zone. In the middle zone the MLD more than 70 t/m² was recorded during periodic weighting.

The figure No. (BRP) 11 shows the pressure distribution between the front & rear legs at different face advance. The higher pressure was mainly confined to mid zone.

Figure No. (BRP) 12 shows the graphs of automatic pressure recorders provided in the powered supports. It may be seen how the pressure changes during different cycles. The setting pressure was 30 MPa and the yield pressure was 40 MPa.

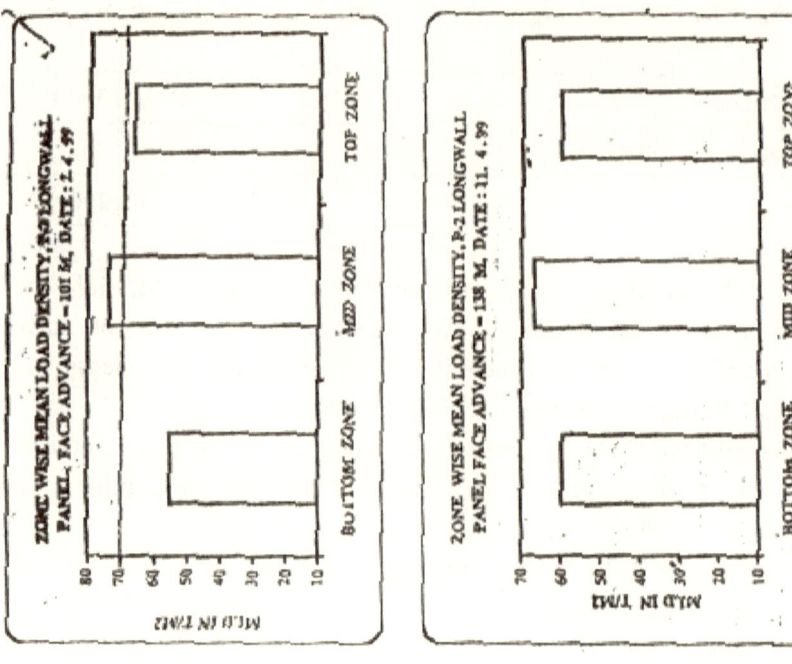

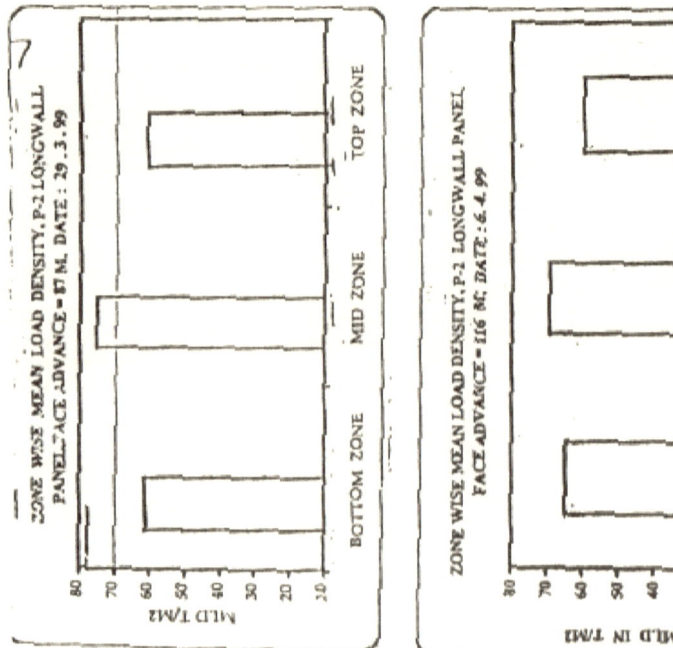

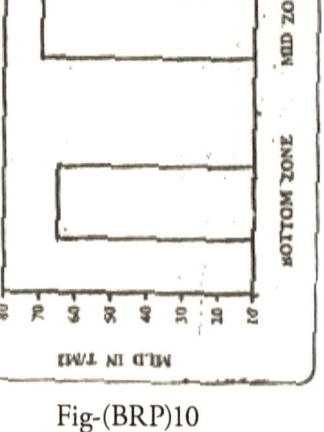

Fig-(BRP)10

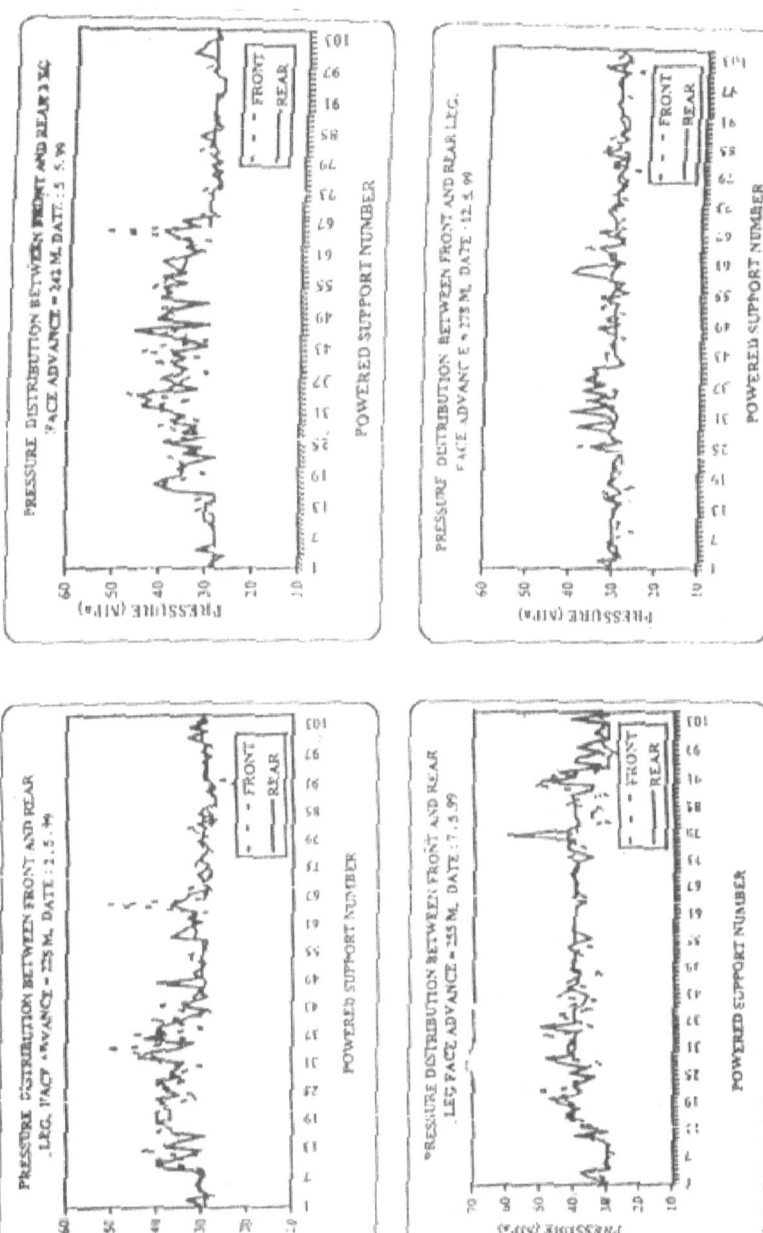

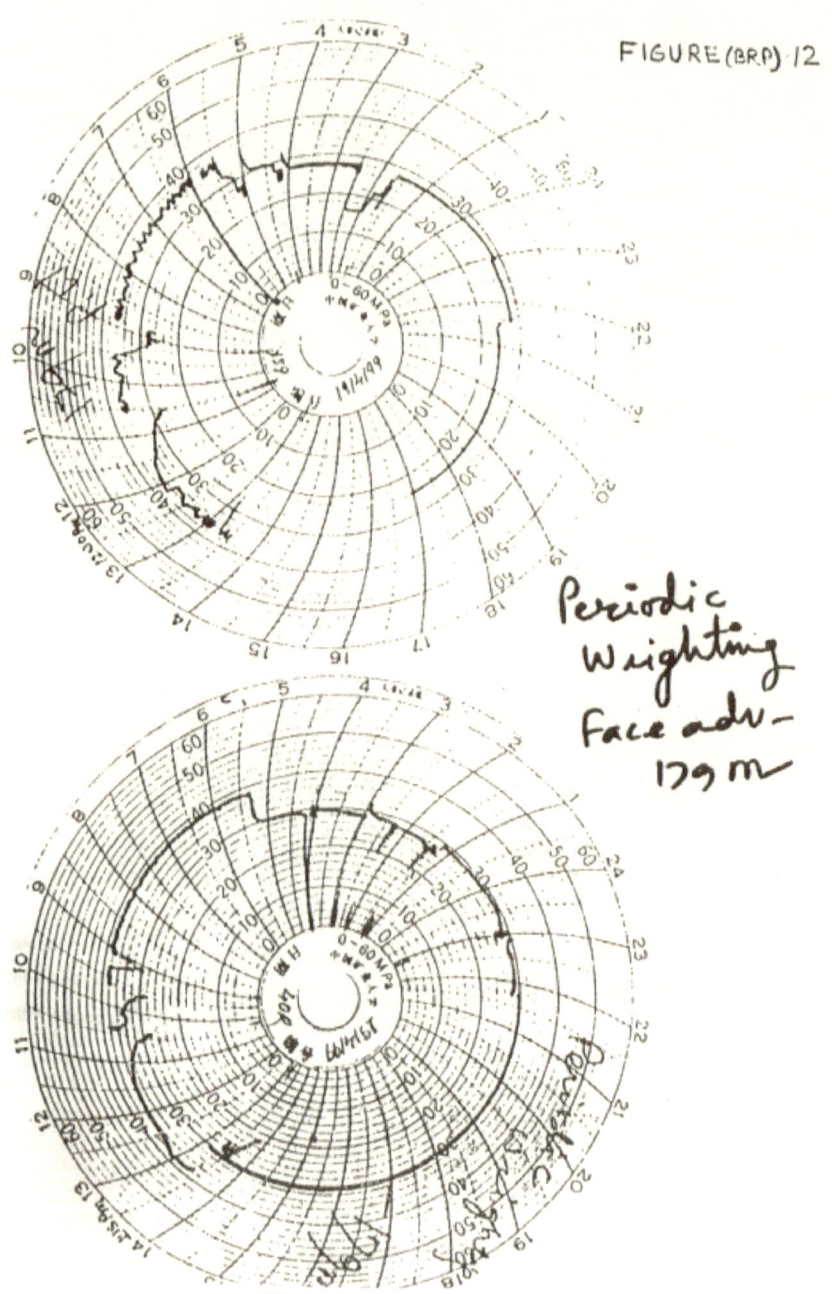

FIGURE (BRP) 12

Periodic
Weighting
Face adv —
179 m

4.2.2.3 IMPACT OF INDUCED CAVING:

The circumstances which led to the adoption of induced caving have been described in the paper earlier under chronological events in Panel P-l. Here the impact of induced caving was reviewed under geological conditions existing in Panel P-2 and the results discussed under the following heads:

(A) Loading on supports.
(B) Periodicity of weighting.
(C) Convergence.
(D) Subsidence.
(E) Goaf filing / packing

Blasting Design:

Blasting was done at 17 m when the face was at a distance of 25 m from the barrier. Second blasting was done at 32 m when the face was at 42 m. Third blasting was done at 47 m when the face was at 58 m. Subsequently double row blasting of special design to create weakness plane was done at 58 m and 62 m. Blasting was done at 77 m. again. Blasting at interval of 15m, 20m & 30m depending on strata characteristics continued up to the end of the panel.

In the blasting, special explosive Sakti prime with maximum quantity of 650 kg for 13 holes were charged. The depth of holes varied from 28 m to 37 m against 54m depth of cover. The vibration in underground near powered supports was measured. In the middle of the face the vibration was from 45.72 mm/sec to 75.5 mm/sec. In the main gate the readings were from 56.9 mm/sec to 69.8 mm/sec. There used to be increase of pressure from 1 MPa to 12 MPas in the automatic pressure gauge recorder.

(A) LOADING ON SUPPORTS:

It was observed that the loading on supports was comparatively less than the loading observed in Panel P-l. In this panel MMLD in 70 % of periodic weightings was at less than 65 t/m^2 and in 30 % cases was between 65-75 t/m^2. So the capacity of the supports with yield load at 79 T/m^2 was found adequate with hydraulic circuits operating at more than 90 % efficiency and induced caving carried out regularly.

The reduced loading on supports might be due to comparatively higher H/t ratio (18) where degree of discontinuity of subsidence trough was less than the panel P-l.

(B) EFFECT ON FREQUENCY OF PERIDOCITY OF WEIGHTING:

Table No (BRP) 17: FREQUENCY OF PERIODIC WEIGHTING:

Interval in M	5-10	10-20	20-30	>30	Total
Frequency	8	23	5	1	37
Percentage	22	62	13	3	100

The above table shown the frequency of interval of weighting and corresponding percentage. In the panel P-2, the blasting was resorted to from the beginning of the panel and the frequency of weighting was 22% between 5-10 m, 62% between 10-20 m, 13% between 20-30 m and rest 3% above 30m.When the figures are compared with Panel P-1 where both blasting and without blasting observations were made, 56% was between 10-20 m, 33% between 20-30m i.e 89% between 10-30 m after blasting compared to 75% in Panel P-2.

It may be concluded that blasting effect had influenced the occurrence of weighting more regularly between 10-20 m and reduction of weighting beyond 30 m interval.

(C) EFFECT OF INDUCED CAVING ON CONVERGENCE:

The convergence recorded during weighting has been discussed later. But the effect of blasting interval when analysed it is found that when the blasting interval was 15 m, cumulative convergence was in the range of 22 mm to 77 mm, when the blasting interval was raised to 20 m the cumulative convergence recorded as (20 to 55)mm, when blasting interval was increased to 30 m, cumulative convergence increased to (94-100)mm. At 60 m interval of blast, cumulative convergence increased to 264 mm. So 20 m interval of blasting was considered suitable for control of cumulative convergence to a safe value for better roof control. However, in order to have effective utilization of explosives the suitable interval between blasting should be further experimented and suitable blast design should be evolved.

(D) INFLUENCE OF INDUCED BLASTING ON SUBSIDENCE:

The induced blasting had influence on angle of draw quite similar to that of Panel P-l. The traveling angle of draw tended to be positive with blasting interval of 15m - 20m and it tended to be negative with the increase of blasting interval.

The subsidence (vertical movement) at a station expressed as percentage of the maximum subsidence (vertical movement) at the same station varied with the variation of blasting interval. It is observed that with 15 m interval of blasting, subsidence at a station was 12 % of maximum on 3rd day after blasting and 60 % on 7th day where as with 20 m interval the corresponding figures were 17 % and 63 %. When the blasting interval was 30m the figures were 11 % and 33 % and with 60 m interval the corresponding figures were 3 % and 7 % From these observation it could be concluded that for regular subsidence blasting interval should be 20 m.

(E) IMPACT OF INDUCED CAVING ON FILLING OF GOAF:

The average RQD at the beginning of the face was 87 %. The RQD during the second main weight at 151 m of face was about 83 % and when the face was crossed 250 m the RQD was reduced to 54 % and subsequently RQD was 43 %.

From the records it was observed that when the face advanced by 5 m of the first instrumented bore hole strata movement started. When the face crossed 10 m beyond the borehole all the anchorages were disturbed. Surface blasting was done at 210m but there was no immediate caving recorded. Second surface blasting was done when the face was 224 m and when it crossed by 25 m, about 2.7 times of thickness of extraction above coal seam caved. The height of caving zone gradually increased to 4.3 times, 6 times, and later on 8 times the thickness of extraction when the face advanced and crossed to about 35 m beyond the borehole position. The RQD near the bore hole was around 80 %.

The surface blasting was done at 285 m before the face reached beneath the second borehole provided with extensometer at 300m from the barrier. When the face reached beneath the borehole fitted with extensometer the last anchorage was lost because of caving. When the face crossed the borehole for about 8 m the caving height extended about 5 times the height of extraction. Subsequently surface blasting at 310 m was done. There was periodic weighting and the height of caved zone extended to about 6 times and then 7.4 times. At that time there was shearer breakdown for 2 days and there was no advance of face and the increase in the height of caving zone was also observed. When the face crossed the borehole for a distance of about 29 m there was periodic weighting and the height of caving extended to 10 times the height of extraction.

From the two records of extensometers it was observed that the strata immediately above the seam was not easily cavable and caving continued in stages due to influence of both the induced caving from surface and also due to weight of the strata. From the records the bulking factor worked out to be about 1.1.

4.2.2.4 CONVERGENCE DURING WEIGHTING:

The first main weighting was observed at a face advance of 87 m and second main weighting at 151 m. Subsequent severe (main) weightings were observed at 383 m and 463 m. High convergence was also observed when the face was at 138m, 179m, 362m. When the face was at 383m very high cumulative convergence was also because of face stoppage due to breakdown and consequent problems in supports.

Table No. (BRP) 18: CONVERGENCE DURING PERIODIC WEIGHTING IN PANEL P-2:

87	major	66	22*(38)	10	First main weight
101		73	60(88)	61	Face stopped for 4 hrs
151	major	259	86*(208)	72	Crack observed in face & cavity formation in roof between PS No. 33 to 63. Second major weight.
171		96	40*(147)		Prolonged Weighting due to maintenance B/D outbye of the face.
179		194	162*(176)		Effect of break down continued from 175 m of face.
204		77	43*(62)		Spalling from face, Roof fall & Water seapage
255		89	37*(75)		Crack on face observed, shale detached from main roof and big pieces of 15 cm thick comes out in mid Zone.
284		29	24*(38)	4.5	
298		18	6(25)	3	
328		6	3(3)		
335		61	25(58)		
350		55	23(42)		
383	major	427	142*(605)	24	Face stopped for 17 hrs. convergence of 363 mm in one cut or 605 mm/ mt was observed. Major convergence during idle hours.
425		95	40(110)		
453		26	14(33)		
463	major	100	42*(80)		
516		20	11(27)		
529		17	14(18)		
560		58	32*(53)		
581		91	51*(65)		No major problems in roof condition
647		20	7(13)		

Following were observed from the table No. (BRP) 18:

(i) The maximum cumulative convergence during the first main weight at 87 m of face was 66 mm and 22 mm/mt of face advance.

(ii) During the major periodic weighting at 151 m the maximum cumulative convergence was 259 mm and the rate was 86 mm/mt of face advance and 72 mm/hr. High convergence resulted in development of crack in the face and roof cavity formation between PS - 33 to 63.

(iii) When the face was at 179 m, the maximum convergence 162 mm/mt was observed but this was because of stoppage of face advance due to breakdown and maintenance continued from the face position at 171 m. The effect of high convergence was reflected in spalling in the face, roof fall and more water seepage.

(iv) Cumulative convergence of around 89 mm and 37 mm/mt of face advance were observed at 255 m when crack on face was observed and a big pieces of shale of 15 cm detached in the mid zone of face.

(v) At 383m, high convergence of 142 mm/mt was recorded as a result of stoppage of face for 17 hours. The convergence of 363 mm was recorded in one cut which is equivalent to 605 mm/mt of face advance.

From the table (BRP) 18, it may be seen that in about 86 % of the cases the convergence rate (mm/mt) was less than 60 and in 67 % of cases less than 40. The higher rate of convergence was observed either during major weighting or during stoppage of face.

Table No. (BRP) 20 shows that in 57% of cases the rate of maximum convergence during cutting was upto 60 mm/mt and in 43% cases it was more than 60mm/mt of face advance.

Although the rate of convergence was more in any particular cutting, the rate of cumulative convergence during cutting was less than 60 mm/mt due to progress of face being maintained in the range of 6m to 7m per day. Whenever, the progress of the face exceeded 9m to 10m, higher convergence along with higher loading was observed.

The high convergence during major weightings and also due to stoppage of face had all resulted in deterioration of roof conditions and problems in shearing/cutting which have been reflected in higher rate of convergence during cutting.

It may be concluded that the critical value of convergence of 60 mm/mt of face advance may be considered for good roof control under this geo-mining condition and if induced caving by surface blasting is optimized both in respect of interval of blasting & blast design the severity of periodic weighting and convergence could be minimized.

Table No. (BRP) 19: Average cumulative convergence in mm/mt of the face advance:

Av. Convergence in (Range mm/ mt of face advance) –Range	0-10	10-20	20-30	30-40	40-60	>60	Total	Remarks
Frequency	3	3	4	4	4	3	21	
Percentage	14	15	19	19	19	14	100	

Table No. (BRP) 20: Max convergence in mm/ mt during individual cut:

Av. Convergence in (Range mm/ mt of face advance) Range	0-10	10-20	20-30	30-40	40-60	>60	Total	Remarks
Frequency	1	1	3	3	4	9	21	
Percentage	5	5	14	14	19	43	100	

4.2.2.5 SUBSIDENCE AND ITS EFFECT ON CAVING:

Subsidence profiles were drawn along the travel of the face and also across the travel. The profile has revealed the typical characteristic of strata movement generally observed at shallow depth of cover. Angle of draw on the dip side was + 8° 41' and on the strike 7°, 40' which established that proper subsidence trough was not developed. Angle of break on the rice side was positive + $10^0 53'$ and on the dip side negative $-21^0 2'$ which may be due to panel barrier effect.

The maximum percentage of subsidence was 42.3 % which was less than the maximum percentage of subsidence observed in Panel P-l. Non-effective width of extraction was recorded as 1.67 Depth.

The ratio of hard cover to height of extraction in the panel was about 18 and the thickness of alluvium & weathered rock was about 14m. The lower percentage of subsidence might be due to better development of subsidence trough and lesser stepped subsidence.

In the panel regular subsidence had helped in fillings up of the goaf and consequently lesser convergence.

4.2.2.6 ROUTINE CONDITION MONITORING:

Regular RCM was carried out by a team. The health of the hydraulic circuits was found to be satisfactory. The unhealthy leg circuits were found to be about 14 % during $1^{H'}$ major weighting. However, during 2nd major weighting the performance of the leg circuits was excellent. Throughout the panel higher efficiency of leg circuits was maintained except on limited occasions.

4.2.2.7 STRATA CONTROL MONITORING IN GATE ROADWAYS:

Gate roadways were supported by 4 rows of 1.5m long roof bolts at lm interval and 2 rows of 40 Tonnes hydraulic props set on either sides. Six load cells were set on top of hydraulic props - 3 nos. in each gate at interval of 10 m covering length upto 30 m from the longwall face. In addition there were mechanical convergence indicators (spring type). The records revealed that the gates were not subjected to additional strata pressure.

4.3 RAJENDRA UNDERGROUND MINE: PANEL P-16

4.3.1 CHRONOLOGICAL EVENTS:

The mine plan showing the location of different panels along with entries is shown in Plan No. (RP) 2.

The face was started with about 2.55 m. height of extraction

Initially it was considered that the overlying strata was easily cavable and therefore no induced caving would be required.

The similar supports of 4 X 513 T (uprated) capacity were used in the face which had 79 T/m support resistance after cut.

The first main weight was observed when the face position was about 77 m from the barrier and periodic weightings at an interval of more than 20 m were observed. When the face was at 103 m there was a major roof fall in the goaf, spalling and crushing sound with water seepage were observed in the face and 3 supports collared due to periodic weighting. This was recovered and the face was started. The face was advanced to 147 m. when 22 supports collared and it took 3 more days to recover the face. When the face was at 164 m there was a periodic weight. Periodic weight at irregular intervals (5 to 28)m continued upto 292m. When the face was at 313 m periodic weight occurred. First 13 supports collared within 3-5 minutes and 30 more supports collared within the next 48-72 hours during recovery operations. The recovery operation was completed and the face was advanced 5m ahead in the stabilized zone. 16 damaged canopies were changed. It took about 18 days to start the face. The longwall operation was started when the face was at 326m. During cutting operation active spalling (10-12)m ahead of the shearer for a distance of about 45 m was observed. The next periodic weight came when the face was at 339m, when the leg closure was 150-250 mm/hr in mid zone. When the face was at 371 m first blasting from surface at 357 m was started. The need for blasting and the impact of blasting on the total operations of the face was studied in detail and it has been discussed later.

After blasting was restored to, the convergence during periodic weighting reduced and regular blasting at different intervals continued upto 937 m and the face continued upto 1002m.

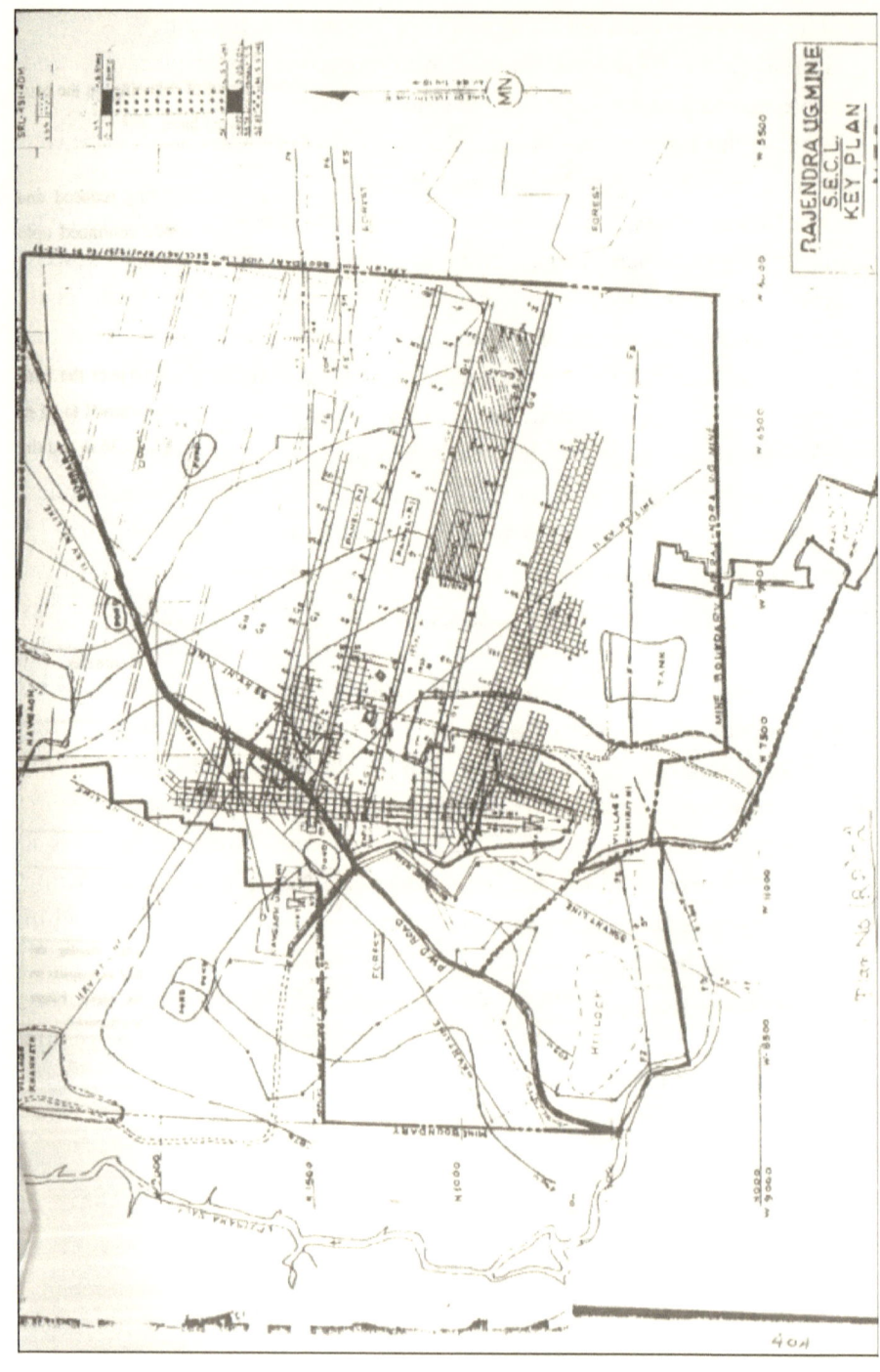

RAJENDRA UG MINE
S.E.C.L.
KEY PLAN

4.3.2 LOADING ON SUPPORTS:

At the start of the panel, thickness of alluvium & weathered rock mantle over the hard, cover was 15m which reduced to 12 m at around 400 m of the panel and increased to 21 m at around 600 m of the panel. The hard stone cover was varying from 34 m to 36 m and the ratio of hard cover to extraction thickness was about 14.

Table (RP) 21 Loading frequency and MMLD during periodic weighting:

	MMLD in T/m²	<65	65-70	70-75	75 to YL	Total	Remarks
Before Blasting	Frequency	81	8	8	20	47	
	Percentage	69	7	7	17	100	
After Blasting	Frequency	81	23	23	42	169	
	Percentage	47	14	14	25	100	
Complete Panel	Frequency	162	31	31	62	286	After blasting the load on supports on the higher ranges had increased.
	Percentage	56	11	11	22	100	

The MMLD before blasting in 69% of the cases was in the range of less than 65 T/m² and in 31% of the cases it was more than 65 T/m². But after blasting the figures were 47 % and 53 % respectively.

The Table No. (RP) 21 shows the loading of supports during periodic weighting in different ranges.

It was observed that the support capacity was inadequate to provide support resistance during major weighting before blasting and after introduction of blasting the measured mean load density reached Rated mean load density in more number of occasions (53 %) but due to lower convergence the face operations could be continued without collaring or damaging of supports.

In comparison with P-2 panel of Balrampur the H/t ratio, average compressive strength and average RQD were lower which might have resulted in formation of discontinuous subsidence trough after adoption of blasting and increased the loading on supports.

FigureNo. (RP) 13 shows the MLD along the face where the maximum loading was in the Mid zone.

4.3.3 CONVERGENCE AND ITS ROLE IN ROOF CONTROL:

The cumulative convergence and rate per hour recorded during main weightings and periodic weightings are presented in table below. The high convergence rate at the beginning had resulted in collaring and damage of the supports.

The troubles of major & periodic weightings at 77m, 103m, 147m and 313m were all associated with collaring of supports. The surface blasting was started at 357 m with face at 371m. During this period (upto 371m) the convergence was high but all could not be recorded. Only in limited number of cases the rate in mm/hr could be recorded.

After introduction of blasting the rate of convergence had reduced remarkably. The rate was high during some occasions due to irregular blasting with increased blasting interval.

Maximum cumulative convergence 500 mm was recorded in Rajcndra P-16 compared to 630 mm in panel P-I and 420 mm in panel P-2 of Balarampur No. 10 & 12 Inclines. In both the cases (P-l & P-2) the blasting was resorted to as a compulsion for improvement of the longwall operation.

In both the panels P-l of Balrampur & P-16 of Rajendra, the hard cover to height of extraction ratios were less compared to P-2 panel of Balrampur. However, the interval of blasting had been playing an important role in cumulative convergence as observed in other panels and here the cumulative convergence increased to 240-345 mm because of irregular blasting interval.

The reduction of convergence even with increased loading on supports resulted in continuance of longwall operation without any major problems of face stoppage and production loss. This was because of better goaf filling.

The rate of convergence in mm/mt of face advance was irregular before blasting but after resorting to blasting, the average convergence was less than 60 mm/mt where no problems in face operation were observed. Likewise convergence in mm/hr during the same period was within 60 mm in 90 % of occurrences. It may be concluded that by induced blasting from surface with suitably designed blast pattern the convergence could be restricted to 60 mm/mt of face advance with support resistance of 79 ton/m^2.

FIGURE (5)-3

PRESSURE VARIATION ON 4280 M
1PM SHIFT AT 5 AM

PRESSURE IN MPa>

POWERED SUPPORTS--->

Table No. (R P)22: Effect of Blasting on Face conv. During periodic weighting:

Face advance	Cumulative convergence in mm	Con. In mm/mt of face advance (in one cut)	Cumm conv. In mm/hr	Remarks
77			80	
103.5			Max (90-120)	3 supports collared
			Avg(-60)	
147		160	Max (2000)	22 supports collared
			Avg . 100	
179		33	Max (120-280)	
			Avg.(-20)	
266	36	20(40)		43 supports collared
313				13 in 3-5 mts.
339			(150-250) Max (200-220) Av (-150)	30 in 48 to 72 m
371	48			First blasting started
396	240	80(183)	8	Blasting interval was 32 m first and them 27 m at 430 m irregular interval
421	345		280	
430	123		40	
445	116		45	
549	144		60	
562	23	13	10	
562	23	13(10)	13	
577	66	37	57	Normal face operation
583	66		34	
637	32	53	53	
640	32		5	
726	67		35	
770	185		70	
802	70	58	83	
813	82		20	
891	288		30	
896	288	96	140	Irregular blasting interval 857 m upto 892 m.
1002	156	52	100	Irregular blasting at 30 m, 8 m & 21 m interval.

4.3.4 INDUCED CAVING – NEED & IMPACT:

From the experience of longwall operations carried out at Balrampur No. 10 & 12 Inclines, it was decided to carry out induced blasting from surface to reduce severity in periodic weighting and control convergence for better roof control. Holes on surface at 56m, 120m, 185m, 315m, 330m were provided but were not blasted at all as these were partly damaged. After severe weighting at 147m face advance when 22 supports were collared and at 313m face advance when 43 supports were collared1, blasting was resorted to from 357m onwards at an interval of 15m upto 680m face advance where 8-10 holes used to be blasted and after 680m upto 837m, 5 holes on an average were being blasted. The interval between blasts was around 14 to 16m (50%) but on some occasion the interval was less than 14m, (28 %) and increased to more than 16m (22 %) as may be seen from the table.

Table No. (RP) 23

Distance/interval in m	<14	14-16	16-20	>20	Total
Frequency	10	18	2	6	36
Percentage	28	50	6	16	100

The distance was being maintained at (1 1-13)m. The details of blasting parameters are as follows:

(a) No. of holes - 5 to 10
(b) Depth of hole - 30 m
(c) Diameter of hole - 100 mm
(d) Distance between consecutive holes in a row - (5 to 7)m
(e) Maximum charge per hole - 50 kg.
(f) Firing system - Nonel (Excel) with N T D.
(g) Type of Explosive

Separate organization was set up with specific responsibilities for ensuring safe **blasting** practices including communication and inspection after blasting.

Records were being maintained. The impact of induced caving has been reviewed under the geological conditions existing in the panel P-16 and the results have been discussed under **the** following heads:

(1) Loading on supports
(2) Periodicity of weightings
(3) Convergence
(4) Subsidence

1. EFFECT OF INDUCED CAVING ON MLD:

Table No. (RP) 21 indicates that before blasting, the MLD was between 65-YL Tonnes/m^2 in 31 % cases and after blasting the loading in this range increased to 54 %. It was concluded that impact of blasting was to increase the MLD at higher range The reasons for increased loading was partly due to lower h/t ratio 14 compared to Panel P-2 of Balrampur No. 10 & 12 Inclines and the alluvium/weathered rock of about 15m to 21m on the top. There was not much disturbance in the longwall except small crushing at the mid zone within 15 minutes to 30 minutes after blasting. There was roof fall and pressure in the supports increased by 2-4 MPa in the mid zone. The caving extended upto the rear end of the support all along the face.

Before blasting, normally spalling of coal from face was observed about 1 to 1.5 cuts before the actual onset of weighting and continued further 5 to 6 cut with severity of throw of coal pieces. But after blasting was resorted to, spalling was limited to 2-3 cuts and throw & size of coal pieces also reduced even though loading on face increased with reduction in convergence.

2. EFFECT OF INDUCED CAVING ON PERIODIC WEIGHTING:

It is observed from the Table No. (RP) 24 below that before the blasting the interval between the periodic weighting was 67% in the range of 10-20m, 28% in the range of 20-30 m and the rest 5% in the range of 5 to 10m. However, after the blasting the interval, between 10-20m increased to 81% and that between 20-30m reduced to 5% only and between 5-10m increased to 14 %. If we take the average of all records in the whole panel, it was 76% between 10-20 m and 13% between 20-30 m, and 11% between 5-10 m.

It may be concluded that periodic weightings between 10-20m had increased occasions after introduction of induced caving from surface.

Table No. (RP) 24:

Range	(5-10) m	(10-20) m	(20-30) m
Before blasting	5%	67%	28%
After blasting	14%	81%	5%
Complete panel	11%	76%	13%

The figure No. (RP) 14 shows the interval of weighting against face position. After blasting the interval between the periodic weightings between (10-20) m was on more number of occasions. The interval on two occasions when the face was at 513.5m and 678m was marginally more than 20m.

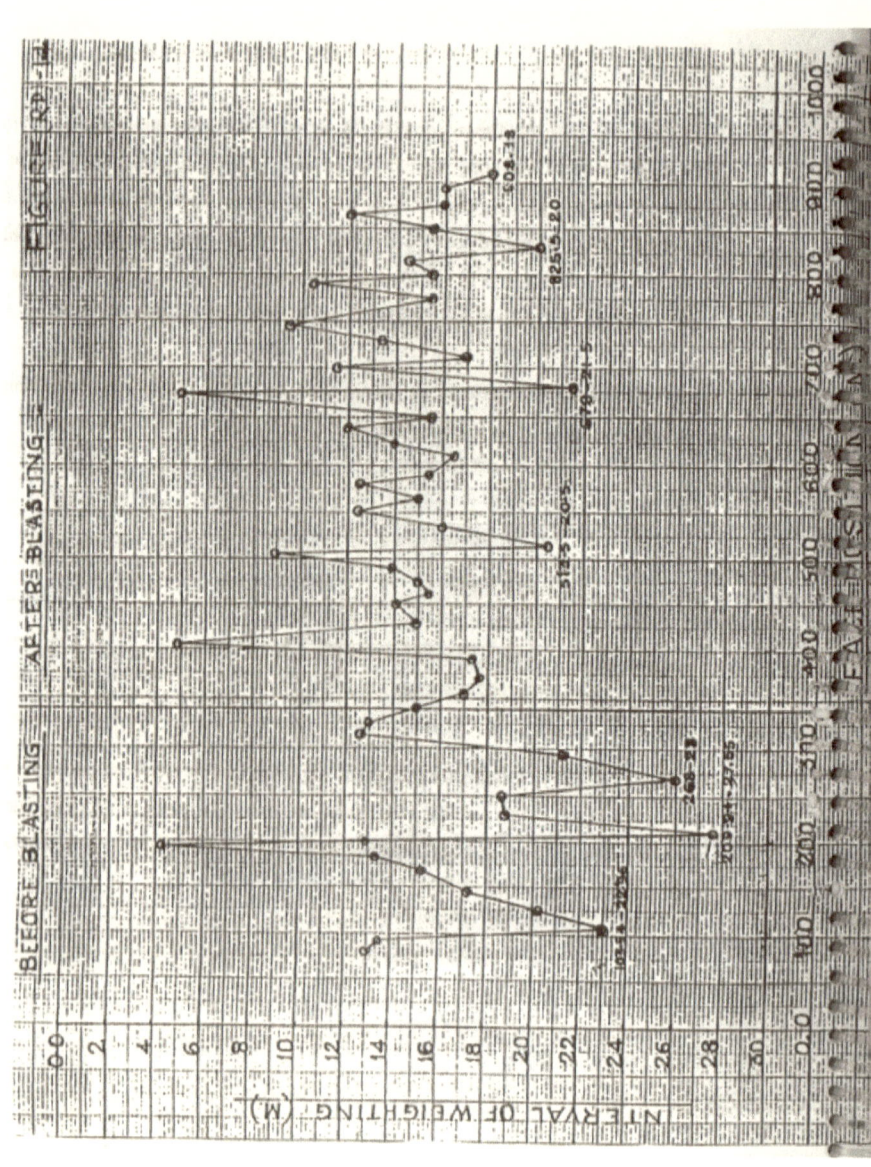

FIGURE No. 1

It may be concluded that the regularity in occurrence of periodic weighting between (10- 20)m increased and the interval more than 20m reduced with more rationalization of blast design & use of explosives, the periodic weights are expected to occur at projected interval.

3. EFFECT OF INDUCED CAVING ON CONVERGENCE:

The parameter of cumulative convergence takes into account both the components i.e. one directly associated with the advance of face and the second one is time dependent component contributed due to stoppage of face for some reason or the other. In this case both the components are presented separately. The time dependent component expressed in mm/hr. would reflect the cumulative convergence over a period of time and high value would mean face stoppage for maintenance / break down etc.

Table No. (RP) 25: Cumulative convergence in mm/mt of face advance:

Convergence mm/m		<40	40-60	60-80	>80	Total
Before Blasting	Frequency	2	Nil	Nil	1	3
	Percentage	67	-	-	33	100
After Blasting	Frequency	2	3	1	1	7
	Percentage	29_	43	14	14	100

The readings of convergence could not be recorded during the initial stage as the face had to be stopped for failure of supports. After the blasting was resorted to, records were maintained but convergence per mt. of face advance was recorded for a limited number of periodic weightings.

The rate upto 60 mm after blasting was for 72 % and rest 28 % when the convergence exceeded 60 mm but the convergence rate in any case had not exceeded 100 mm.

*Table No. (RP) 26: **Cumulative Convergence in mm/hr.***

Convergence mm/hr		<40	40-60	60-80	>80	Total
Before Blasting	Frequency	1	1	1	2	5
	Percentage	20	20	20	40	100
After Blasting	Frequency	16	3	1	1	21
	Percentage	76_	14	5	5	100

Before blasting rate of convergence upto 60 mm/hr during periodic weighting was in **40 %** cases and above 60 mm it was 60 %. But after blasting, the convergence upto **60** mm increased to 90 % of cases. The high rate in cases before the blasting established high idle time convergence. The rate of advance of face when maintained at 6-7 m per day the convergence was restricted to 60 mm/m and the roof control problems were manageable but when rate increased to 9m/day it resulted in higher loading which could have resulted in collapse of supports.

The high rate of convergence in one case even after blasting was because of irregular blasting with_increase Of blast Interval to 33 m but the impact of induced caving in general was reduction of convergence. This had been reflected in both the components.

4. EFFECT OF INDUCED CAVING ON SUBSIDENCE:

The overlying strata did not subside as the extraction was completed. Initially the immediate roof was falling/breaking filling up the goaf partly but these were not reaching the surface The induced caving had accelerated fall of the next overlying strata The subsidence of any particular point expressed as a percentage of the total subsidence of the point indicate how the strata movement was taking place. The table (RP) 27 below shows the subsidence percent before and after blasting on different days counted from the undisturbed day.

Table No. (RP) 27:

Days from the undisturbed day	3 rd day	9 th day	14 th day
Before blasting	3.7 to 13.5	83 to 96	98 to 100
After blasting	50-70	84 to 99	95 to 100

It may be seen that there was good difference in vertical fall of a station on 3rd day but subsequently there was not much difference in the vertical movement (subsidence). It may be concluded that induced caving accelerated the development of subsidence on surface at the beginning and subsequently the movement of strata was gradual till it reaches the full subsidence.

4.3.5 INFLUENCE OF STRATA IN CAVING:

From the records of borehole extensometer it is observed that when the longwall face crossed the instrumented borehole, there was no movement of anchorage. When the face crossed by 6/7 meters, the movement started at the lower anchorages & when the face crossed about 25/26m, the caving reached upto 4t (where t is the thickness of extraction in m) and when the face reached about 30m ahead, the caving zone reached upto 13t. In case of lower average RQD (less than 50), the caving started earlier and the caving zone reached upto 7t. The bulking factor thus worked out to be 1.08 to 1.14 depending on the type of strata.

4.3.6 SUBSIDENCE AND ITS EFFECT ON ROOF CONTROL:

The maximum subsidence was recorded as 53.3 % which is higher compared to subsidence observed in Panel P-2 of Balrampur Colliery. The difference in subsidence was probably due to difference in Hard Cover to thickness of extraction ratio. The ratios in Panel P-2 and P-16 were 18 & 14 respectively as a result discontinuous subsidence trough associated with stepped subsidence resulted in comparatively higher percentage of subsidence.

The angle of draw on the dip side was recorded as + 6 Deg. which was lower than the value observed in Panel P-l of Balrampur mine.

Regular subsidence observations at the shallow depth of cover had helped in assessing the extent of overhang underground and subsequent decision on action plan for suitable blast design. Regular caving and filling up of goaf resulted in lower convergence improving the roof condition.

4.3.7 STRATA CONTROL MONITORING IN GATE ROADWAYS:

Load cells were set on the 40 Tonnes hydraulic props at 10 m interval upto 30 m from the face in both the gate roadways. The spring loaded fixed wire convergence meter at 10 m interval along with load cells were provided in both the gate roadways.

Records of measurements indicated that at the beginning of the longwall face upto 354m, there was an increase of load upto 1 to 1.5 Tonnes when the face reached within (4 to 5)m of measurement station in the gate roadways. After the blasting was resorted to, no such increase of loading was observed. No convergence was recorded both before and after surface blasting was resorted to.

5.0 DISCUSSIONS

5.1 LOAD ON SUPPORTS AND SUPPORTS CAPACITY:

In Panel P-l, initially the support resistance was 69 T/m^2 which was uprated to 79 T/m^2. In Panel P-2 & P-l6, the support resistance was 79 T/m^2 as the support was uprated before start of panel.

During operation of panel P-l of Balrampur and Panel P-l 6 of Rajendra the Measured Mean Load Density (MMLD) during major weighting reached the rated Mean Load Density (RMLD) of 69 T/m^2 to 79 T/m^2 respectively associated with high convergence of 126 mm/minute & 100 mm/minute respectively resulting in collaring and damaging of supports and partial failure of faces. This was probably due to block movement of the roof in the near vicinity of the extraction horizon resulting in sudden additional pressure, higher rate of convergence and failure of face.

Because of presence of strong bed within 9/10 times the thickness of extraction, the caving was not complete. As the extraction was being carried out at shallow depth of cover, the strata weight was not adequate to break the strata and reduce overhang. As a result during main weighting the support resistance was not adequate and the supports collapsed. Introduction of induced caving increased the MLD as the H/t ratio was made discontinuous due to induced break. In panel P-l & P-16 the H/t ratios were 11 & 14 respectively and in P-2 he ratio was 18. In Panels P-l & P-16, the loading intensity was higher and the MLD reached the Yield load where as in P-2 the MLD even during major weighting did not reach the Yield load. If we consider the efficiency of the support as 70 % due to use in number of panels, the rated support resistance should be in the range of 105 to 110 T/m^2. So even with induced caving the support capacity should be such as to provide 105 to 110 T/m^2 support resistance.

The original & revised support capacity recommended by CMRI based on the *"Experience of longwall operations at shallow depth of cover"* with H/t ratio varying form 7.4 to 11 at Raniganj Coalfield was found to be inadequate under the conditions existing in both the mines. Higher capacity of support was needed in these mines under South Eastern Coalfields Limited.

5.2 CONVERGENCE AND RATE OF ADVANCE:

In Panel P-l and P-16 after partial failure of faces, induced caving was introduced. In Panel P-2 induced caving was resorted to from the beginning and the longwall operations could be completed without any major problems. The induced caving had direct impact in reduction of cumulative convergence, convergence in mm per meter of face advance and convergence per hour. Although the MLD in both P-l and P-16 increased due to low H/t ratio, the reduction in convergence had contributed in completion of all the panels with the same capacity support without experiencing any major strata control problems.

The cumulative convergence had two components first contributed by advance of face and second one was time dependent contributed by stoppage of face i.e. idle time. From the records, the rate of face advance proved to have direct influence on the convergence and higher rate contributed to roof control problems. The average rate of advance of 6/7 meter/day had made it possible to control the convergence rate and advance beyond 9m resulted in increase of convergence and loading on supports due to delayed caving and consequent increase of overhang. The second part of time-dependent convergence was found to be high when the face advance was stopped due to breakdown etc.

5.3 INDUCED CAVING FROM SURFACE:

The load measurements at all the longwall faces before the major weighting resulting in partial failure of face at P-l and P-16 clearly indicated that during the periods other than weighting, load on face and supports was less than the designed capacity of the support system and the roof condition was satisfactory.

However during periods of major weighting the maximum mean load density (MMLD) in the mid /one of the face reached the rated mean load density (RMLD) and the supports collapsed within a short period of time and supports were damaged. The rate of convergence during these periods was also high. This phenomenon prompted the management and the Director-General of Mines Safety to think on some solution to continue the operation with same capacity of supports as higher capacity of supports was not immediately available. The immediate solution of induced. caving along with uprating of the supports to design limit was thought of and execution planned. In the Panel P-2 the induced caving was resorted to from the beginning of the panel and strata control measurements were continued. In the panel P-1 and P-16 after the recovery of faces, the blasting was resorted to and the faces were started. The strata control measurements continued. The strata control measurements from all the faces have established that induced caving had resulted in increased load on supports but the rate of convergence and cumulative convergence had reduced and face could be continued without any major strata control problems. The blast design covered different parameters like design of holes, charging interval of blasts, type of explosives etc and proved to be an important operation which needed scientific planning and execution by a team of experts.

In the paper the impact of induced caving on periodicity of weightings, convergence, loading on supports, subsidence and caving behaviour was discussed separately for each panel. It may be concluded that the induced caving had increased loading on supports, reduced convergence, increased periodicity of weighting between 10-20m and reduced periodicity beyond 30m. The induced caving had also increased the initial subsidence by stepping and reduced the negative traveling angle of break. The blasting had increased the height of caving zone after the face crossed the point for a distance of about 30m. The blasting interval of 20m was established to be an appropriate interval for better roof control and induced caving to be continued for successful operation of Longwall face so long the 4- legged 4X513 capacity powered supports are in use.

However experiments on suitable blast design and optimization of use of explosives are to be carried out for further improvement In longwall operations.

5.4 SUBSIDENCE AND ITS EFFECT ON LONGWALL OPERATION:

The longwall operations at Panel P-l, P-2 and P-l6 all were carried out at shallow depth of cover with H/t ratio 11, 18 & 14 respectively. At shallow depth of cover the subsidence trough becomes discontinuous and it occurs in two forms i.e. stepped subsidence and pot holing. In longwall operation at shallow depth of cover planning should ensure that pot holing does not occur. In the longwall panels at Jhanjra with H/t ratio varying from 7.4 to 11 pot holing occurred only in one panel with H/t ratio of 9.7 reported to be due to geologically disturbed area associated with long stoppage of face due to break down etc. The records at three panels of mines under SECL indicated that at lower H/t ratio (11) loading was comparatively more and the subsidence was high. For future extraction it may be concluded that extraction at H/t ratio less than 15 should not be planned to guard against pot holing & higher stepped subsidence. In all Extraction at shallow depth of cover where discontinuous subsidence trough is likely to develop, support capacity should be suitably augmented to provide adequate support resistance for increased loading compared to longwall mining at higher depth of cover (more than 100m) where regular subsidence trough would be formed. The subsidence profile along the travel of the longwall face showed undulating profile with irregular humps at places where maximum subsidence was expected to be continued.

5.5 EFFICIENCY OF HYDRAULIC SYSTEM SUCCESS OF OPERATION OF LONGWALL:

In panel P-l initially the efficiency of support system was not satisfactory i.e. unhealthy circuits were more than 10 %. After the major weight at 160m the support components O were examined thoroughly and defects were identified and rectified. The support capacity was uprated to 40 MPa and it was decided that the setting load should be around 30 MPa.

In subsequent operations in all the panels (P-l, P-2 & p-16) the setting load was maintained around 30 MPa. The installation of automatic pressure recorders provided continuous guidance in ascertaining the support load and consequent action needed.

Routine Condition Monitoring (RCM) was introduced and it was continued in all the panels and it was considered that the RCM had helped in identifying the defects and executing corrective measures which in turn made it possible to maintain more than 90 % of hydraulic, circuits in healthy condition practically throughout the panel.

5.6 STRATA CONTROL MONITORING:

Records generated by observations in underground at longwall face, gate roadways and in surface by subsidence survey and instrumented borehole survey by a team of the management provided lot of information & guidance in proper control of longwall operation

PANELWISE BREAKDOWN ANALYSIS

Fig-(BRP) 15

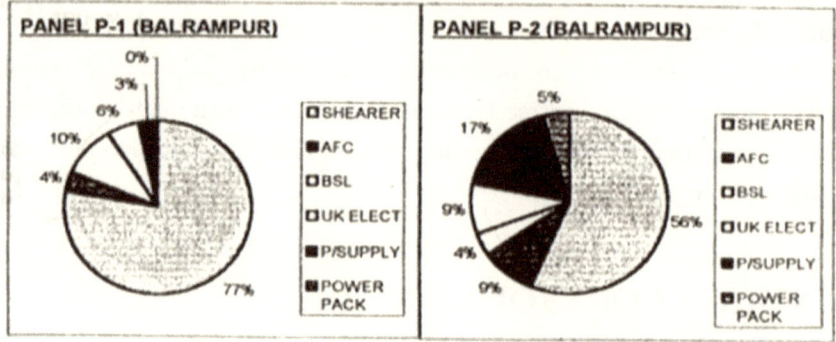

PANEL P-1 (BALRAMPUR)

0%
3%
6%
10%
4%
77%

- ☐ SHEARER
- ■ AFC
- ☐ BSL
- ☐ UK ELECT
- ■ P/SUPPLY
- ■ POWER PACK

PANEL P-2 (BALRAMPUR)

5%
17%
9%
4%
9%
56%

- ☐ SHEARER
- ■ AFC
- ☐ BSL
- ☐ UK ELECT
- ■ P/SUPPLY
- ■ POWER PACK

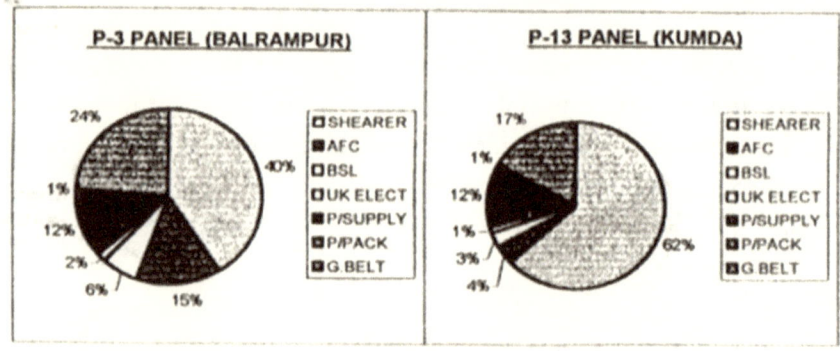

P-3 PANEL (BALRAMPUR)

24%
40%
1%
12%
2%
6%
15%

- ☐ SHEARER
- ■ AFC
- ☐ BSL
- ☐ UK ELECT
- ■ P/SUPPLY
- ■ P/PACK
- ■ G.BELT

P-13 PANEL (KUMDA)

17%
1%
12%
1%
3%
4%
62%

- ☐ SHEARER
- ■ AFC
- ☐ BSL
- ☐ UK ELECT
- ■ P/SUPPLY
- ■ P/PACK
- ■ G BELT

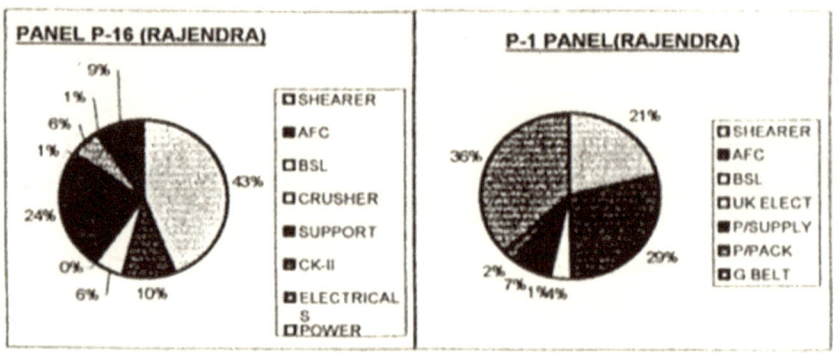

PANEL P-16 (RAJENDRA)

9%
1%
6%
1%
24%
0%
6%
10%
43%

- ☐ SHEARER
- ■ AFC
- ☐ BSL
- ☐ CRUSHER
- ■ SUPPORT
- ■ CK-II
- ■ ELECTRICALS
- ☐ POWER

P-1 PANEL(RAJENDRA)

21%
36%
29%
2%
7%
1%
4%

- ☐ SHEARER
- ■ AFC
- ☐ BSL
- ☐ UK ELECT
- ■ P/SUPPLY
- ■ P/PACK
- ☐ G BELT

BSL - Stage Loader.

DELAY ANALYSIS OF PSLW PANELS

Fig-(BRP)16

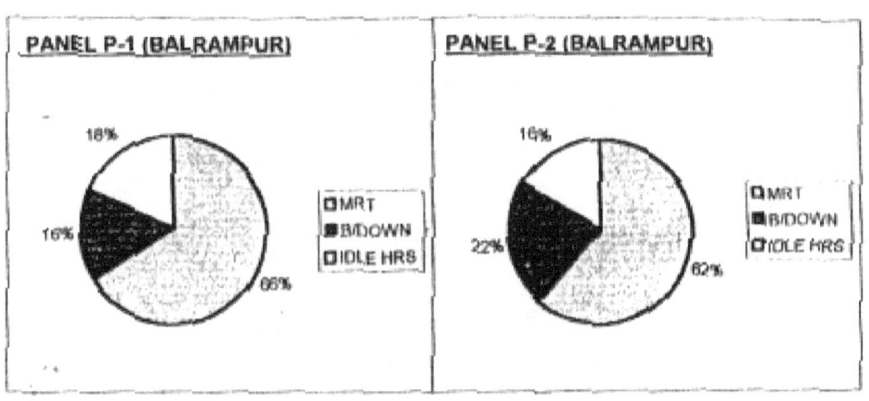

PANEL P-1 (BALRAMPUR)
18%
16%
66%
☐ MRT
■ B/DOWN
☐ IDLE HRS

PANEL P-2 (BALRAMPUR)
16%
22%
62%
☐ MRT
■ B/DOWN
☐ IDLE HRS

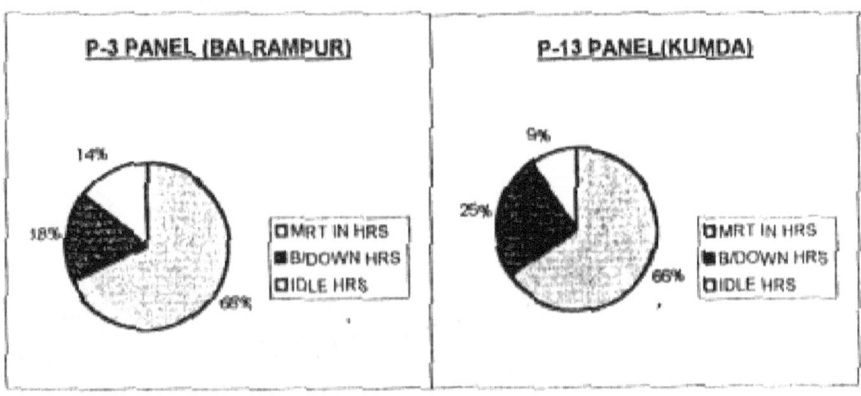

P-3 PANEL (BALRAMPUR)
14%
18%
68%
☐ MRT IN HRS
■ B/DOWN HRS
☐ IDLE HRS

P-13 PANEL (KUMDA)
9%
25%
66%
☐ MRT IN HRS
■ B/DOWN HRS
☐ IDLE HRS

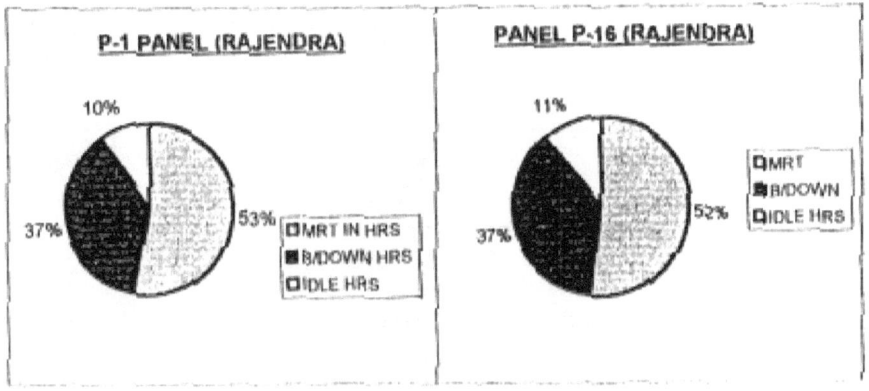

P-1 PANEL (RAJENDRA)
10%
37%
53%
☐ MRT IN HRS
■ B/DOWN HRS
☐ IDLE HRS

PANEL P-16 (RAJENDRA)
11%
37%
52%
☐ MRT
■ B/DOWN
☐ IDLE HRS

6.0 MANAGEMENT OF MEN & MACHINES

6.1 OPERATIONAL PROBLEMS

Under the present day scenario, reliability and capacity utilization of the equipment are guiding factor for any capital investment. From the experience gained by the operation of PSLW face equipment in the past it revealed that out-bye clearance, reliable power supply and delays in face preparation were the few main reasons for low machine running time.

The Table- 28 placed below shows the delay analysis of 19 faces of C1L.

TABLE-28

MRT IN % OF MAT	MAT LOST DUE TO BREAK-DOWN IN %					
	Shearer	AFC/ STL	Gate belt	Power pack/ chock	Elect	Total
30.50	8.7	9.48	3.65	3.53	3.75	29.1

MAT LOST DUE TO DELAY IN %					
Shift change	Geo –min condition	Power failure	Face preparation	Out –by clearance	Total
3.74	9.65	2.44	16.16	8.39	40.4

From the above it can be seen that delays on account of above was 69.49% of the machine available time. Proper precautions were taken to take care of those elements before introduction of PSLW face in the mines of SECL. SECL has so far operated and completed 2 panels at Balrampur and one panel at Rajendra. Table-29 placed below will show the delay analysis in the panel of the above 3 panels. Pie diagrams showing panel wise breakdown & delay analysis in respect of all the pslw panels already worked or being worked are shown in figure no 15 & 16.

Table-29

PANEL NO.	MRT in % of MAT						
		Shearer	AFC/ STL	Gate belt	Power pack/ chock	Elect	Total
p-1	63%	14.2	1.9	1.9	0.6	.8	19.4
p-2	62%	12.6	2.7	2.0	4.6	2	23.9
p-16	52%	15.8	5.3	2.9	3.9	2.7	30.6

	MAT LOST DUE TO DELAY IN %					
	Shift change	Geo –min condition	Power failure	Face preparation	Out –by clearance	Total
p-1	Nil	11.65	1.3	Nil	4.7	40.4
p-2	Nil	5.8	3.7	Nil	4.6	
p-16	5.3	4.6	2.8	Nil	4.7	

While the total time loss due to breakdown & delays could be reduced from approx. 70 % in 80s to 37 % now, lots of problems were experienced during operations of those panels particularly due to breakdown of shearers & powered supports.

The reliability of the Shearer was low. Downtime of Shearer alone was 46% to 82% of the total break-down. This was mainly due to failure of the casing of ranging arms of the Shearer. Detailed analysis through finite element analysis revealed that factor of safety particularly in the region near the out-put shaft of the drum was low. All the six casings of the ranging arm failed. Subsequently, the ranging arms have been modified with superior design and appropriate selection of metallurgy.

In order to assess the real time condition of the powered support, pressure gauges were fitted in each leg circuit of the powered support. The pressure gauges could read the instantaneous as well as peak kick pressure of the individual legs. In addition to above, 10 nos. of continuous pressure recorders were fitted in the powered support at different locations all along the face to know the strata behavior in terms of load on the support. The monitoring of the pressures of the supports were being done continuously.

6.2 PRODUCTIVITY & PROFITABILITY

Detailed analysis of financial result will reveal that one of the major reasons for loss from the underground mines is low man productivity. While the EMS has increased substantially over the years, the underground man productivity has remained practically constant at 0.57 - 0 59 for CIL. Wage cost alone is approx 80% of selling price of Grade 'D' coal. Since installation of PSLW face at Balrampur, it has completed 2 panels and extraction of the 3rd panel is in progress. Rajendra has also completed 1st panel and operation in Panel No.2 is continuing.

The Table 30 & 31 placed below will show the production and financial results:

TABLE-30

S.NO	Parameter	Panel no.2(Balrampur)	Panel no. 16 (Rajendra)
	PRODUCTION	444016 T	622221 T
1	Wages	113.90	148.93
2	Overheads	43.76	43.92
3	Stores	80.90	33.26
4	Power	74.51	61.18
5	Coal transportation	40.82	48.52
6	Description	72.67	52.63
7	Interest	47.11	133.70
8	Welfare	5.15	10.02
9	Miscellaneous	7.74	11.85
Cost of production		469.08	502.73
Quality adjustment		(-) 8.14	(-) 6.76
Sale value		839.49	907.77
Net sale value		831.30	901.01
Profit		362.22	398.28

A comparative study has been made on financial results for the Rajendra mine when the mine was run by conventional Bord &pillar vis- a vis powered support.

TABLE-31 Rajendra u/g Mine

S.No	Parameters	97-98	96-97	95-96	98-99	Remarks
1	Production in (L.T)	1.78	1.63	1.60	4.87	PSLW started in Dec. 98 & worked
2	OMS	.76	.78	.75	1.92	For 4 months and
3	EMS (in RS.)	391.95	379.33	207.51	420	Prodn. 2.92 Lakh T
4	Loss of prod(Rs/T)	1129.97	925.42	770.13	597.92	
5	Sale price (Rs/T)	892	915.51	696.77	835.19	
6	Loss/ profit (Rs/T)	(-) 237.97	(-)9.92	(-)73.63	237.27	

7.0 RECOMMENDATIONS

(1) The Longwall at shallow depth of cover (less than 100 m) may be planned but support capacity should be suitably augmented to take care of additional loading due to discontinuous subsidence trough. The minimum support resistance should be 105 to 110 T/m^2 for extraction of thickness of 2.5m.

(2) The minimum depth of cover for successful operation of Longwall should take into account the formation of discontinuous subsidence trough and pot holing. In the mines under SECL the hard cover to extraction thickness ratio should not be less than 15.

(3) The induced caving from surface where resorted to should be planned by a team of experts for suitable blast design and optimisation of use of explosives.

(4) The average rate of advance of face should be maintained at 6/7m per day for control of convergence to a limit of 60 mm/mt of face advance

(5) Routine Condition Monitoring and Strata Control Monitoring shall be planned and implemented by the management to develop confidence and expertise amongst the people who are responsible for success of longwall operation.

(6) More scientific studies should to be conducted to develop a mathematical / numerical $0r$ model to assess the support capacity & cavability based on the geo-technical and other parameters influencing the operations of longwall.

(7) To ensure higher reliability of the equipment, continuous flow of spares are to be ensured.

(8) On line health monitoring of the support to be developed to know real time health of the supports.

(9) Till the time on line health monitoring of supports is developed, each leg circuit of powered supports to be provided with pressure gauges & continuous pressure recorder at each 5th support in the mid zone & 10th support in the main & tail gate zones.

(10) There is an urgent need of development of indigenous spares as well as equipment to improve the machine available time.

(11) Technology of induced caving by long hole blasting from surface is a great break through in hard roof management at shallow depth of cover. This should be suitably modified for hard roof management at higher depth also.

8.0 FUTURE SCOPE OF APPLICATION

In view of the above, this technology of Longwall Mining with induced caving by deep hole blasting particularly at shallow depth of cover, could be gainfully utilised in the virgin properties of Shivani mine, Rehar, Gayatri, Karkati, Damini, Amba & Baduli under South Eastern Coal Fields of Coal India Ltd.

Moreover, 1200 million Tonnes of coal is blocked under standing pillars in different seams and extraction of these coal seams could be done economically with safety by Mechanised Short Long wall with Deep Hole Blasting from surface by use of powered supports of adequate capacity. However, the past experience of crossing galleries with parallel face should provide guidance in designing line of face with appropriate alignment with the developed galleries/junctions. Adequate consideration should also be taken to restrict the H/t ratio to guard against formation of pot holes at the gallery junctions.

9.0 ACKNOWLEDGMENT

The authors would like to express their Sincere thanks to Shri S N Padhi, Director-General of Mines Safety, Ministry of Labour, Government of India & Shri G K Jha, Chairman-cum- Managing Director, M/s South Eastern Coalfields Limited for their support to present the paper in the Seminar. The contribution made by the following officers are also thankfully acknowledged: S/Shri.

KM Prasad,	Director of Mines Safety
P K Sarkar,	Director of Mines Safety,
Rahul Guha,	Director of Mines Safety(R & D),
BP Singh,	Deputy Directors of Mines Safety
T K Mandal,	Deputy Directors of Mines Safety
G Vijaya Kumar	Deputy Directors of Mines Safety
K Chadda	General Manager, Bishrampur area SECL
KDSood,	General Manager, Shohagpur area, SECL.
N Prasad-	Project Officer/Agent, Balarampur Project.
S.Shrivastava-	Manager of Balarampur Project.
N Das	Project Officer/Agent, Rajendra Project.
B Pandey	Manager of Rajendra Project.
A K Saxena	Sr.M.E.Central Mine Planning and Design Institute.

Thanks also expressed to the officers and staff contributed in the preparation and presentation of the paper.

BIBLIOGRAPHY

(1) Shri S.R.Mehta- "Experience on Long wall Technology at Jhanjra Project" presented in a Seminar on High Production
Technology for Underground Mine organized by MGMI on 12th &13th July, 1996.

(2) Shri T.K.Mozumdar- "Subsidence in Indian Coal Mines", M.Tech. Thesis submitted in Indian School of Mines in 1984.

(3) Dr. S.K.Sarkar- "Mechanized Long wall Mining, the Indian Experience" presented in Second National Conference on Strata Control in India.

AB-189, SALT LAKE,
SECTOR-I. KOLKATA-64
PHONE: 033-2334- 4467
Mob: 09748567628
E-mail: tkmozumdar2@gmail.com

CURRICULUM VITAE
T.K.MOZUMDAR
DECEMBER 2014

He obtained B.Sc.(Mining Eng.) in 1964 and M.Tech (Mining Eng.) in 1984 from IndianSchool of Mines, DHANBAD.

He worked in the Directorate General of Mines Safety (DGMS), GOI from 1966 and retired as Deputy Director General of Mines Safety in June2000.

He was awarded Gold Medal by Mining, Geological and Metallurgical Institute(MGMI) in 2000 for the contribution in successful operations of Powered Support Operated Long Wall (PSLW) Method at shallow depth of cover by induced caving.

He was appointed in 2003 as consultant in a Joint Project of the U.S. Department of Labor and Ministry of Labour, GOI, on "COAL MINE SAFETY AND HEALTH PROJECT" and completed reports on Health and Safety Management Plans of 5 Indian Coal Mines

He is presently attached to PEIL, Kolkata as Consultant on "Improvement of Health and Safety in Mines, Steel Plants and other Industries."

He had more than 40 publications in International and National Level Seminars/Conferences.

F-E-143, SEC-III
SALTLAKE, KOLKATA-700106

Mob: - 9830024733
biswanathpan1943@gmail.com

CURRICULUM VITAE
BISWANATH PAN

Born in 1943 and graduated in Electrical Engineering in 1965 from B.E College, West Bengal.

He joined Bengal Coal Company, in 1965 and later on served in Coal India Ltd. He worked as Chief GM from 1993-1997, worked as Director (Tech/Operation) from 1998-2001 and as Chairman-Cum-MD from 2001 and superannuated in 2003.

He was extensively trained on Underground Mine Mechanization and in maintenance, repair and major overhauling of PSLW equipment from 1981-1987 by Dowty, Gullick Dobson of UK, Kopex of Poland and Germany.

He is a life member of Mining Geological and Metallurgical Institute (MGMI) of India.

> 1999-2000- **He was awarded Gold Medal by MGMI in 2000 for the contribution in successful operations of Powered Support Operated Long Wall (PSLW) Method at shallow depth of cover by induced caving with blasting from surface.**
> 2003-2004. He was awarded **Sukumar Raksit Medal by MGMI** for his outstanding contribution in mine mechanization.
> 2008-2009: He was also awarded the **Engineering Gold medal by MGMI** for his innovative contribution in the underground mining.

After his retirement he was the pioneer in successful introduction of High Wall Mining Technology by use of Superior High Wall Miner for extraction of thin seams as low as 0.9m and multiple seams.

He is now working as CEO cum DIRECTOR of two private companies.

He has presented more than 50 papers on mine mechanization and various techniques in the national & international conferences/seminars.